S0-FEO-659

F/NF
15.00

# FREE ENERGY AFLOAT

# FREE

# ENERGY AFLOAT

by
Nan and Kevin Jeffrey

SEVEN SEAS PRESS, INC.  *Newport, R.I.*

Published by Seven Seas Press, Inc., Newport, Rhode Island, 02840

Copyright © 1985 by Kevin and Nan Jeffrey
Art © by Kevin Jeffrey

All rights reserved. No part of this book may be reproduced in any form or by any electronic or mechanical means, including information storage and retrieval systems, without permission in writing from the publisher.

NOTE: The information presented in this book is accurate and complete to the best knowledge of Seven Seas Press at time of publication. Alternate energy equipment, design, engineering, construction and pricing are subject to change at any time. Due to the nature of the marketplace and the potential for widely divergent use of alternate energy equipment and technologies, the authors and publisher disclaim any liability incurred by the use of information and equipment described and pictured in this book.

**Library of Congress Cataloging in Publication Data**

Jeffrey, Nan, 1949-
   Free energy afloat.

   Includes index.
   1. Boats and boating—Electric equipment.
2. Renewable energy sources.  I. Jeffrey, Kevin,
1954-      . II. Title.
VN325.J44  1985      623.8′503      85-2187
ISBN 0-915160-78-1

1  3  5  7  9   RR/RR   8  6  4  2  0

Designed by Irving Perkins Associates
Printed in the United States of America
Edited by James R. Gilbert
Technical editing by David Mac Lean

*To Jim Kirby, a long-time friend, sometime sailor, and one of the "old guard" alternate energy enthusiasts . . . for keeping faith over the years.*

# ACKNOWLEDGMENTS

Our sincere thanks to the manufacturers of marine alternate energy equipment that provided help and information for this book, with special thanks to Jack Condos, Hamilton Ferris, Steve Gaunt, Hugh Merewether, and Bill Owra for their patience and enthusiasm; to the Donut Shoppe of Vero Beach, Florida, for providing "office space" and sustenance during the early stages of this project; and to all of the sailors who welcomed us aboard, plied us with refreshment, and imparted information and anecdotes about their alternate energy systems.

# CONTENTS

PROLOGUE: Consciousness, Commitment, Conservation   xvii

## PART I—Laying the Groundwork   1

### CHAPTER 1: Understanding Energy   3
- *What is energy?*   3
- *How does it work?*   3
- *Sources of energy*   4

### CHAPTER 2: Basic Electricity   8
- *Taking out the mystery*   8
- *Some simple analogies*   8
- *Electrical concepts*   9
- *Instruments*   10
- *Types of electricity*   13
- *A simple 12V circuit*   14
- *Batteries*   14
- *A boat's electrical schematic*   22
- *Safety*   24

### CHAPTER 3: Marine Alternate Energy Equipment   26
- *Solar systems*   26
- *Wind systems*   26
- *Water systems*   39

## PART II—Assessing Your Needs   43

### CHAPTER 4: How Electrical Appliances Determine Load   45
- *Reducing your electrical demand*   45

- *A table of typical marine appliance loads   46*
    - cabin lighting   48
    - anchor lights   49
    - running lights   51
    - refrigeration   51
    - disposable batteries   55
- *Electrical load of four typical boats   57*

CHAPTER 5:  Selecting An Alternate Energy System   59
- *For a dinghy   59*
- *For a daysailer   61*
- *For a boat smaller than 28 feet   61*
- *For a boat larger than 28 feet   62*
    - sloops   62
    - cutters   62
    - ketch/yawl/schooner   63
    - catboats   63
- *General hull characteristics   64*
    - multihulls   64
    - design characteristics   64
    - ancillary equipment   65
    - boat image   66

CHAPTER 6:  Boat Use And How It Affects Your Decision   67
- *The weekend sailor   67*
- *The occasional-cruise sailor   67*
- *The voyaging/live-aboard sailor   68*
- *The working live-aboard sailor   68*
- *The ocean-crossing sailor   69*
- *Sailing with children, pets and friends   70*
- *The charterboat owner   70*
- *The racer   71*
- *The offshore racer   72*
- *Making a deadline   72*

CHAPTER 7:  The Importance of Climate   74
- *Solar radiation   74*
    - in the U.S.   74
    - outside the U.S.   74
    - micro climates   76
    - mini-micro climates   76

- *Prevailing winds   77*
    - coastal U.S. winds   77
    - micro climate   77
    - mini-micro climate   78
- *Water   79*

CHAPTER 8:  Hybrid Systems and Other Alternatives   81
- *A wind/water hybrid system   81*
- *A water/solar hybrid system   82*
- *A wind/solar hybrid system   83*
- *A complete hybrid system   83*
- *Supplementing AE systems   84*
    - portable generators   84
    - inverters   86
    - boosting engine alternator output   87

# PART III—SELECTING, INSTALLING AND OPERATING YOUR SYSTEM   91

CHAPTER 9:  Matching A System To Your Needs   93
- *Notes on sizing your AE system   93*
- *Estimating your system's daily output   94*
    - solar   95
    - wind   95
    - water   96
- *Four typical installations   96*
    - AE for boats consuming 6 amp-hours a day   96
    - AE for boats consuming 14 amp-hours a day   97
    - AE for boats consuming 40 amp-hours a day   97
    - AE for boats consuming 100 amp-hours a day   97

CHAPTER 10:  Other Alternate Energy System Components   98
- *Batteries   98*
- *Wiring/terminals   99*
    - wire type   99
    - connectors   100
    - terminals   100
- *Diodes   101*

- *Instruments/Meters   101*
    - ammeter   102
    - voltmeter   102
    - battery condition indicator   102
    - anemometer   102
    - knotmeter   103
- *Voltage regulator   103*
- *Automatic load disconnect   104*
- *Emergency shutdown switch   104*
- *Fuse/circuit breaker   104*

CHAPTER 11: Owner-Built Alternate Energy Systems   105
- *Building your own solar panel   105*
    - installation   106
    - reflectors   107
    - trim   107
- *Making your own wind generator   107*
    - rigging-suspended mounts   108
    - pole mounts   110
    - fixed mounts   110
- *Making your own water generator   111*
    - trailing log type   111
    - auxiliary generator on freewheeling shaft   112

CHAPTER 12: Installing The System   113
- *Solar panels   113*
    - deck mount   114
    - quick disconnect   115
    - stern mount   116
    - general notes   117
- *Wind generators   117*
    - pole mounts   117
    - masthead mounts   119
    - rigging-suspended units   119
    - general notes   120
- *Water generators   120*
    - trailing log type   120
    - auxiliary prop through the hull   120
    - auxiliary generator on freewheeling shaft   121
    - general notes   121

**CHAPTER 13:** Operating And Maintaining Your System   122
- *Solar systems*   *122*
- *Wind systems*   *123*
    - self-tending units   123
    - non-self-tending units   123
    - general notes   125
- *Water systems*   *127*
    - trailing log type   127
    - auxiliary prop through the hull   128
    - auxiliary generator on freewheeling shaft   128
    - general notes   129

**CHAPTER 14:** Taking Your System Ashore   130

# PART IV—REFERENCE SECTION   133

**CHAPTER 15:** A Glossary of Electrical Terms   135

**CHAPTER 16:** A Little Theory   143
- *Photovoltaics*   *143*
    - light   143
    - solar cells   144
    - solar panels   145
    - notes on solar panel output   147
- *Wind generators*   *148*
    - notes on wind generator output   150
- *Water generators*   *151*
    - notes on water generator output   153
- *Generators/Alternators*   *154*
    - permanent magnet DC generators   154
    - alternators   156

**CHAPTER 17:** The Marketplace   159
- *Solar panels*   *159*
- *Wind generators*   *170*
- *Water generators*   *183*

**CHAPTER 18:** Comparing Alternate Energy Systems   191
- *Performance—a product ranking*   *191*
    - solar panels—all conditions   191
    - wind generators—light wind   192
    - wind generators—strong winds   192
    - wind generators—general notes   192
    - water generators—general notes   193

- *Ease of operation   193*
    solar panels   193
    wind generators   193
    water generators   194
- *Reliability/Lifespan   194*
    solar panels   194
    wind/water generators   194
- *Safety: a system-by-system ranking   195*
- *Appearance: the authors' ranking   196*
- *Size and weight   197*
    solar panels   197
    wind/water generators   197
- *Noise/Vibration   197*
    solar panels   198
    water generators   198
    wind generators   198
- *Maintenance: a system-by-system ranking   198*
- *Ease of repair   199*
    wind generators   199
    water generators   199

# FOREWORD

This book is a milestone for sailors that will engender a lot of excitement about the usefulness of renewable energy at sea. This opinion is based on considerable direct experience.

In the fall of 1982, a new type of fishing vessel was launched. It was a 32-foot Ocean Pickup trimaran, designed and built by Dick Newick. The vessel was designed as a fisheries research boat, particularly adapted to the needs of Third World fishermen. The boat was named *Edith Muma,* and was pressed into immediate testing that first winter trawling in Nantucket Sound.

In the spring, the *Edith Muma* made a 4,000-mile voyage to Georgetown in Guyana, carrying close to a ton of research gear and supplies. I was aboard for the leg from Bermuda to South America. It was an exciting trip, dominated by tumultuous tropical skies, periodic storms and the timelessness of space created by the sea. We were almost alone, making contact only twice with other ships.

The Ocean Pickup in many ways is unique. It is fast, seaworthy and is an excellent fishing platform. It may also be the first research vessel and likely the first fishing vessel to be exclusively supported by photovoltaics. The *Edith Muma*'s log, auto pilot, running lights, VHF radio, compass and depth recorder/printer all are powered by the sun.

Those people who believe workable alternate energy systems are still in the future may be surprised to learn the *Edith Muma*'s photovoltaic system never has given us any trouble. On the trip to Guyana I became a firm convert to the notion of energy independence at sea, the subject of Nan and Kevin Jeffrey's book *Free Energy Afloat*.

In Guyana we discovered that the *Edith Muma* could pay for itself in the offshore gill net fishery on fuel savings alone in one year, thanks not only to the wind but to the sun, as well. In the winter of 1984, Dick Newick and I sailed the *Edith Muma* up to Tobago and then over the Spanish Main to Puerto Viejo in Costa Rica. There she remains, with my son Jonathan aboard as captain. The vessel has opened up new fisheries, stabilized the local fish supply and attracted regular fish buyers to this remote community. The Ocean Pickup often works around the clock three shifts a day, with more than half the village fishermen participating. She has made money for the villagers and converts to a modern age of sail. Ironically, she has also brought solar power to a community that up to now has not enjoyed all the fruits of 20th century technology. When the boat's solar panel is underutilized, it charges local truck and radio batteries for our neighbors.

*Free Energy Afloat* has inspired me to become more ambitious with regard to renewable energy. We are developing new Ocean Pickup fishing vessels with refrigerated seawater holds and electric haulers. Our plan is to operate this equipment off hybrid renewable energy systems. The improved value of the catch will pay for these improvements. In some places, it may be necessary to use fossil fuel engines as a backup. But in many places, especially the trades, the entire task will be handled by the wind, currents and the sun.

With *Free Energy Afloat* the Jeffreys have provided a step-by-step guide to alternate energy for the uninitiated, and a variable manual of options for the expert. Its striving for an energy independence afloat will provide lessons for rethinking how we use energy on land. Through the Jeffreys and their book, we can see how boats literally are the vessel of change towards a coming solar age.

JOHN TODD, PH.D.
President, Ocean Arks International
Falmouth, Massachusetts

PROLOGUE

# THE FUN AND ELEGANCE OF ALTERNATE ENERGY

The term "alternate energy" suffers from the same ambiguity as the word "natural". What does it really mean? Different people use different meanings to suit their own needs. A natural cereal in the grocery store might contain three types of sugar, two kinds of salt, and a hydrogenated vegetable oil. But by the use of the word "natural" the label implies "healthful" to the unwary buyer. Similarly, the Department of Energy customarily defines "alternate energy" as any alternative to fossil fuels, including nuclear energy and use of synthetic fossil fuels. In our opinion, they have stretched a term with a strong environmental impact to include environmentally unsound sources of energy.

For our own purposes we will use the definition favored by most environmentalists and "soft" energy proponents. "Alternate energy" comprises energy sources that are presently created *entirely* by the sun. This includes our complete list of renewable energy sources in Chapter 1, which are all by-products of the sun's energy and occur in various forms over the entire surface of the Earth. They do not have to rely on large, centralized stations to refine, convert, or distribute them, and are usually most economical when applied on a small, local scale. They depend on relatively low technology levels and are labor-intensive, creating jobs. Their ecological impact is small, especially if they are used to power many small facilities rather than fewer, large ones. They can be, if desired, independent of utility companies, giving individuals considerable control over the price and availability of the energy they need. If this country totally adopted the use of alternate energy it could achieve a political stability, economic security, and environmental harmony not experienced since the industrial revolution. If this is true, why aren't we committed to this technology?

There are three major reasons for man's reluctance to accept alternate energy. First, many people look at one energy source, like solar, and say "solar can't take care of our energy needs, so alternate energy must not be feasible." It's true that no *one* alternate energy source can satisfy all our energy needs the way oil, coal or nuclear energy have, but by using *all* of them, by diversifying and localizing the energy sources, we believe we can do it.

Second, there is a widespread belief that alternate energy technology is still in the experimental stage. The bane of alternate energy enthusiasts is someone standing there looking at a solar panel or wind generator and saying, "it's certainly the thing of the future." People seem unwilling to accept alternate energy at anything less than perfection, yet they buy cars without a thought that one year down the road their particular model will be refined and improved and thus become obsolete. The same applies to household appliances, furnaces, outboards, electronics, computers—in fact every mechanism they buy without hesitation. So why is it so hard for them to believe tht alternate energy technology is here and available now?

Finally, people cling stubbornly to a tunnel-vision way of comparing the cost of alternate energy equipment only with today's retail *cost of fuel*. But where do we take into account the cost of cleaning up the environment? Who pays for the lakes and streams dying from acid rain caused by the widespread burning of sulfur-laden coal, the "greenhouse" effect in our atmosphere from carbon dioxide (a by-product of combustion), the oil spills, the nuclear plant clean-ups and dismantlings after mere 20-year lifespans? What type of economic indicator can you place on leaking, spent radioactive fuel or the scarring of the earth by strip mining? The cost of these clean-ups is so vast it is almost incomprehensible. The money will only come from one place, our taxes. What does it cost this country that a major part of our foreign policy is determined by our need to keep our oil interests intact? Quite obviously there are certain things that have not been entered into our economic considerations and others that simply cannot. There is no cost/benefit ratio that can justify results of our energy blindness as drawn in National Geographic's Energy Report: "Over the long sweep of history, human beings will look back and note with awe (and chagrin) that their ancestors stripped the planet of most of their exhaustible endowment within the span of a few hundred years. Twentieth-century people alone will have used up the bulk of it."

The rapid diminution of fossil fuels is forcing us to choose between a "soft" renewable energy path and a "hard" non-renewable one of nuclear power and synthetic fuels. Surprisingly, even though we feel that the decision really shouldn't be made purely on economic grounds, the soft-energy path has proven economically more sound. Amory Lovins, nuclear physicist and leading authority on how all types of energy production and use affect the world both economically and ecologically, has shown in his book *Soft Energy Paths* that by increasing our end-use efficiency and practicing conservation, we will find that the soft energy path is faster, easier, less expensive and much more socially attractive than our current one.

Now that we know where to go, how do we get there? It's easy. Three things can make this transition smooth: consciousness, commitment and conservation.

## CONSCIOUSNESS

Become aware of the total energy picture, both now and in the future. Don't look at just the present retail price of fuel but at the additional economic, environmental and social costs as well.

*Prologue: Consciousness, Commitment, Conservation*

## COMMITMENT

More than anything else, the success of alternate energy depends on a state of mind. Time and again throughout history we have seen the tremendous ability of a nation committed to one goal. When President Carter came to office his energy advisors predicted that alternate energy would supply 5% of this country's energy needs by the year 2000. Four years later they were predicting 20%. What happened? Did the sun increase its heat output? Did the wind suddenly blow harder? Of course not. All that changed was a frame of mind, a desire to go after 20%. Given time (but no actual change in the technology) they might have said 40%. When you realize that our lack of effort is not a state of physics but a state of mind, you see what can be achieved by commitment.

## CONSERVATION

There are three ways in which we can effectively conserve energy. First, we can develop and use more energy–efficient appliances, tools, transportation, buildings, etc. Second, we can look at our end–use efficiency and focus on energy forms that require few changes or none at all. Most of all we need to develop an awareness of how our personal energy consumption affects the world around us. Even renewable resources can be ecologically destructive if we abuse them. Think of your own use of energy as similar to the practice in Iceland of taking down from the nests of the eider duck to make down quilts and clothes. Only as much down is removed as the birds can safely life without. A perfect balance has been reached enabling both the animals and the industry to thrive. We need to establish a similar balance between our energy consumption and the world that supplies that energy.

Sailors are an excellent group to convert to alternate energy to supply *all* their electrical needs. They are already using vehicles that provide clean, efficient and reliable transportation from that renewable resource, the wind. They are also separated from the conventional electrical grid. In the past sailors have depended solely on auxiliary engines to supply electricity, an inefficient process that takes the *chemical* energy of liquid gas or diesel (oil derivatives) to fuel internal combustion engines that rotate a *mechanical* shaft that turns an electrical generator that sends *electrical* energy to be stored as *chemical* energy in a battery. Obviously, the end-use efficiency of such a process is less than ideal, although it is slightly higher than land electrical power plants, which suffer losses in transmission over power lines. In addition, it is costly to both fuel and maintain an engine for this purpose. Engine generation is only efficient when the boat is under power and therefore underway, for then the energy produced by the engine is accomplishing two jobs at once. This is called *cogeneration,* or the production of electricity as a by-product of some other process. Industry is just now waking up and realizing the benefits of this kind of energy recycling.

The elegance of alternate energy for sailors becomes evident whenever they're on their boats. Engine-driven energy generators are noisy. Alternate energy sources are quiet. The weight and space taken up by extra fuel is far bulkier than those required by a wind generator or solar panel. Traditional methods take something simple (electricity) and make it dependent on something complex (a marine engine). With alternatives you have only a small generator to maintain and repair, something well within the capabilities of most sailors. Gas and diesel are only available at marinas or populated areas. Alternate energy is available everywhere. Engine fuel is polluting, non-renewable and costly. Sun, wind and water are clean, renewable and free.

By now you should have a feel for the elegance of alternate energy but where, you are probably wondering, is the fun? What's so fun about maneuvering around a solar panel (something else to clutter up the decks) or raising a wind generator at the end of a long day's sail (what, another job?). And what about the problem of the water generator fouling the taffrail log or fishing line already trailing off the stern? You call that fun? Well, we never said it would be effortless, but the minor difficulties can be ironed out. Sailors have already learned to live with minimal headroom, confined spaces, a tippy home and shattered schedules. Raising a wind generator or juggling a solar panel is going to seem like child's play compared to tying in reefs or struggling with anchors. The fun comes from watching all that free energy pour into your batteries and out your lights or radio or VHF. It's fun every time your neighbor cranks on his engine as you thank God you don't have to do *that* anymore. As with most converts, those who embrace alternate energy tend to adopt it as a hobby. Alternate energy buffs can constantly be caught flicking the little switch that activates the battery condition indicator. During a blow we practically trip over each other trying to be the first one to the ammeter to read out "over 10 amps!" Anyone who has lived with alternate energy will probably tell you it's fun.

Another fun aspect of alternate energy is its ability to turn a potentially uncomfortable situation into a more enjoyable one. When a cold front comes roaring through your previously idyllic anchorage, it's consoling to know your wind generator is churning out electricity. Instead of sitting inside bemoaning the weather, we turn on the lights, start up the tape deck, and enjoy the surplus energy. On hot, windless days as you lie wilting under the Bimini, you think kind thoughts about your solar panel. And when an extra-long sail keeps you underway longer than desired, it's comforting to look back and see that water generator putting in overtime offer the stern.

Some purists might argue that if simplicity and self-sufficiency are the goal, then kerosene lights are the answer. Unquestionably, kerosene is simpler than alternate energy for it avoids electricity altogether. On our boat our two Tilley kerosene lanterns often light and warm the cabins on cold nights and are a good backup in case the battery gets low. But kerosene still involves work, is a non-renewable resource, costs money and can be hazardous, especially with children on board. Electricity is pleasant, easy, odorless, and most people's first choice. Alternate energy is electricity at its simplest. With it you can enjoy any of today's wonderful gadgets, from lights to electronics.

For the sailor, especially, the ability to generate electricity independent of land and fuel docks is the greatest reward of alternate energy.

*Kevin and Nan Jeffrey*
*Barnstable, MA*
*June, 1985*

# PART I
# Laying the Ground Work

# CHAPTER 1

# UNDERSTANDING ENERGY

Energy is a simple word with a dynamic impact. Throughout history it has been profoundly involved in the development of civilization. Primitive man used only the energy of his own muscles until he discovered fire, his first alternate source of energy.

Today, when we speak of energy, we mean energy external to our bodies. This is the energy that provides the backbone of our civilization—comfortable habitats, agriculture, machines, transportation and most recently, space travel. Without it we would still be living primitively; with it we have developed our highly-industrialized society. But the price has been great.

The bulk of our energy comes from non-renewable sources. That is, we mine it from the Earth, we don't grow it or take it from an essentially infinite reservoir. As it is used up, it becomes more expensive.

## WHAT IS ENERGY?

The dictionary defines energy as the capacity to do work and overcome resistance. This chapter will take you a step further towards a better understanding of how energy works and what forms it can take.

Energy exists in either mechanical, electrical or chemical forms or in the form of heat or radiation. It may be in motion (*kinetic*) or at rest (*potential*). An example of kinetic motion is the energy of a waterfall or the electrical current lighting a bulb. An example of potential is the energy contained in the water behind a dam, or the charge stored in a battery. Kinetic energy is in the act of being used while potential energy is being held in reserve.

## HOW DOES IT WORK?

In addition, energy can be converted from one form to another. Each time it changes form the total amount of energy stays the same. This is known as the Law of Conservation of Energy. The *usable* energy, however, is always less. Think of this gradual loss of usable energy as similar to cashing a traveler's check in a foreign country. Each time you change currency you receive less than the face amount. Each time you convert energy to a new form you lose some in the process.

The conversion of a gallon of oil into electricity for your home is a good example. At a power plant, a gallon of oil (stored chemical energy) is burned (changed into heat energy), the heat boils water and turns it into steam, the steam turns turbines (mechanical energy) that then turn electrical generators (electrical energy). The electricity moves to your home (with line and transformer losses) where it is used to heat your home (kinetic heat energy), run your dishwasher (kinetic mechanical energy), or operate your lights (visible radiation energy). The energy lost in the process of energy conversions usually takes the form of heat, which is very hard to retrieve and use. It's easy to see that electricity generated from oil takes a big loss in changing forms, losing two-thirds of the original usable energy in the conversion processes. The less you change form, the more usable energy remains, or the greater the *end-use* efficiency.

In contrast to generating electricity from oil or coal, the human body provides an excellent example of efficient energy use. We eat food that our body converts into usable chemical energy to power our basic functions. This is clearly the energy of our dictionary definition; the ability to do work and overcome resistance. Some of the energy is stored as potential chemical energy to be used on demand by our central nervous system that sends electrical impulses to our muscles. Our muscles are then used to run, walk or work (mechanical energy). The energy lost in these conversions is heat energy that warms our body and keeps us at a steady and very critical 98.6°F. Beautiful! Food energy is *internal* to our body. But there is a physical limit to this internal energy. Man, being hard to please, now wants energy to perform great amounts of work and overcome vast resistances. Consequently, he has become increasingly dependent on *external* energy.

## ENERGY SOURCES NOW AVAILABLE

External energy falls into two categories, renewable and non-renewable. First, we will discuss the renewable energy sources that all, in one way or another, result from sunlight hitting the Earth:

### Solar Energy

What we think of as solar energy is actually the result of a self-sustaining nuclear fusion reaction of a star, 93,000,000 miles from Earth, which we call our sun. The energy released is constant, even though the usable energy reaching the Earth is affected by clouds and smog and the tilt and rotation of the planet.

When the rays of the sun, which consist of radiant heat and light energy, reach the Earth, they are converted to sensible heat (heat we can feel and measure) by every object that absorbs them. They also can be changed directly into electrical energy by the photovoltaic cell. As a point of interest, today's cell was first developed for the space industry, although silicon cells have been around since the 1930's.

### Wind Energy

Wind, which is the air around us in motion, carries mechanical energy. The motion is caused by variations in temperature due to non-uniform solar heating of the Earth's atmosphere, which is why wind is considered another form of solar energy. The rotation of

*Laying the Groundwork*

the Earth determines the basic flow patterns of the wind. As sailors are aware, sailboats use the wind's energy directly to provide momentum through the water. Windmills use it to grind grain, pump water or turn machinery. It also can rotate propellers that drive electrical generators to create electrical energy.

## Hydroelectric Energy

Still another form of solar power is hydroelectric energy, which is the mechanical energy of moving water converted to electricity. The sun evaporates water from the surface of the Earth. The water then falls as rain or snow. In high places the rain or melted snow comes under the influences of gravity and flows down until it reaches the ocean, its state of lowest potential energy. By using the natural flow of some rivers, this moving water can be harnessed to drive turbines (mechanical energy) that rotate generators (electrical energy). Unlike a hydroelectric plant that uses the force of moving water to generate electricity, a water-powered electrical generator for boats uses the motion of the boat to pull a propeller through the relatively motionless water.

## Biomass

Biomass contains stored chemical energy and consists of anything organic, such as wood, organic wastes, plants, peat, etc. Biomass produces usable energy three ways. It can be converted to heat by direct combustion (burning wood or peat), it can be fermented and then distilled to produce the liquid fuel ethyl alcohol (or ethanol), and organic wastes can be converted into methane gas fuel (anaerobic digestion). As an additional bonus, all forms of biomass energy can be produced from surplus, or otherwise unusable, organic matter.

## Tidal and Wave Energy

Tidal and wave energy is mechanical energy created by the movement (daily in some places, twice daily in others) of the tides, and by wind energy converted to wave motion as the wind blows across the water. Electrical generators can harness this wave energy, converting it into electricity. Technically, this type of energy derives from a combination of moon power and solar power.

## OTEC—Ocean Thermal Energy Conversion

This term is applied to a system that uses the substantial difference in temperature between the surface of the water that is warmed by the sun, and the deeper, colder water below. Greatly simplified, the steps in the process are as follows: The warm water vaporizes a low-boiling liquid (transfers its energy to the liquid), which drives a turbine-generator to produce electricity. During this process, the vapor loses much of its own heat energy. It is then further cooled by means of the cold deeper water, so that it condenses back to the liquid state, ready for a new cycle.

This system operates best in the tropics where the surface water is warmest.

### Geothermal Energy

Geothermal energy is heat energy stored in the Earth. The radioactive decay of rocks heats the Earth below the surface, creating a heat energy source, which can create steam that can be used directly to heat homes and buildings or to drive turbines to generate electricity.

## NON-RENEWABLE ENERGY SOURCES

As mentioned earlier, all energy sources discussed so far are renewable provided we use them with discretion. The sun will continue to shine (at least until a very distant future), and cause the wind to blow, waves to roll, rivers to flow, and organic matter to grow. Our second energy category is comprised of the non-renewable types, which spawned the industrial revolution and created our present standard of living. This vast wealth of energy comes from only three sources: hydrocarbon fossil fuels, nuclear fission and synthetic fuels.

### Hydrocarbon Fossil Fuels

Fossil fuels contain potential chemical energy created by organic matter that was trapped in the Earth millions of years ago and converted into oil, coal or gas by intense pressure and temperature. Forty-two gallons of crude oil are equivalent in energy content to 5,700 cubic feet of natural gas, or ¼ ton of coal.

*Oil* is the liquid form of fossil fuel. It is first refined, then either burned directly, yielding energy to heat our homes or to drive turbine generators that produce electricity, or burned in the internal combustion engine to create mechanical movement, as in automobiles and other machines.

*Coal* is fossil fuel in solid form. Although coal was originally used for the same purposes as oil, its use is now mostly limited to coal-fired electrical power plants. It's hard to extract from the ground and costly to transport. When coal or oil burns, sulfur is released into the air as sulfur dioxide. Different types of fossil fuels contain different amounts of energy and sulfur

*Natural gas* is the gaseous form of fossil fuel. It is easy to transport, is comparatively clean and has the same basic uses as oil and coal. It is either burned directly to provide heat and cooking fuel or used to convert heat energy to steam to power mechanical equipment.

### Nuclear Fission

Nuclear fission produces energy by breaking up enriched uranium atoms to produce a large amount of heat. Enriched uranium is uranium that has been subjected to a sophisticated process that makes it "fissionable" or "fissile". A pound of ordinary uranium yields almost three million times the energy supplied by a pound of coal. A special form of nuclear reactor—the breeder reactor—converts some of the uranium to plutonium while it is undergoing ordinary fission. Plutonium is also an effective fissile fuel. The reactor, therefore, "breeds" one type of fuel while consuming another.

### Synthetic Fossil Fuels

Synthetic fossil fuels convert some solid sources of energy to a more usable, transportable form. Three examples of synthetic fossil fuels are petroleum from oil shale rock formations, gas from coal (coal gasifaction), and liquid fuel from coal (liquefaction).

We have now taken a look at what energy is, how it works, what sources are presently available, and what uses they can be put to. Next we will examine some of the principles of electrical energy in order to better understand alternate energy systems for boats.

# CHAPTER 2

# BASIC ELECTRICITY

## INTRODUCTION: TAKING OUT THE MYSTERY

In Chapter 1 we talked about how energy exists in several forms, either mechanical, electrical, chemical, heat or radiation. Here in Chapter 2 we are concerned with electrical energy and the way it behaves. Many people probably don't understand electrical energy because they can't see it. We can see the effect of electricity in a light bulb, or hear its effect when a radio is turned on, or feel its effect when it operates a fan. We understand the work being done but not the electricity itself.

Several things can help take the mystery out of electricity. First, electricity is very predictable. A few simple rules and relationships tell us where, when and how strong (or useful to us) electricity is at a given moment. They also will tell us where to look if electricity is not where we thought it should be.

Second, several inexpensive instruments, when used correctly, help make electricity "visible" to us.

## ANALOGIES

### Electricity/Water Analogy

The flow of electricity is often likened to the flow of water, which, unlike electricity, is something we can see. The analogy is as follows:

**Fig. 2-A.** Water/electricity analogy.

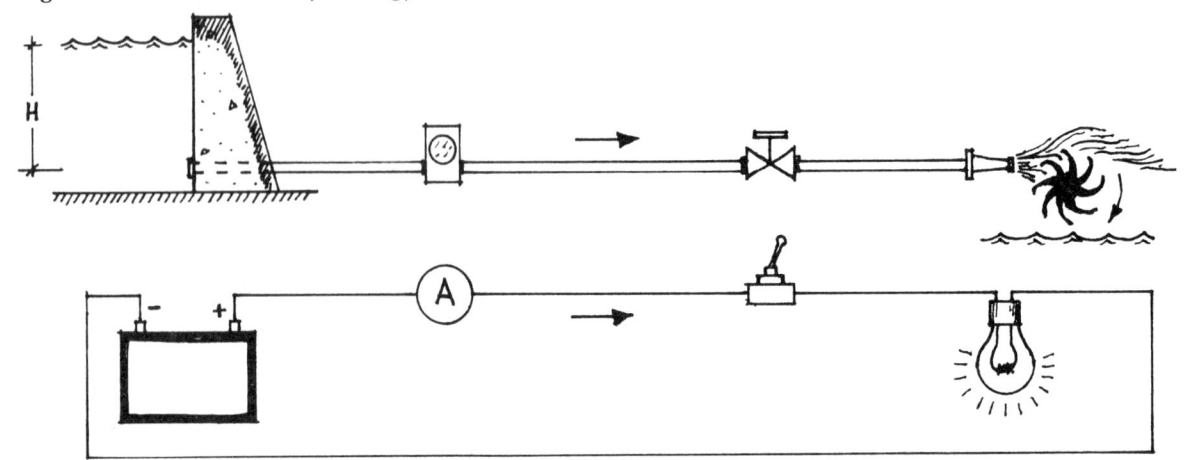

A reservoir stores mechanical energy as potential energy in the form of water pressure, which is directly proportional to water quantity and height. When we want to use that pressure, say to turn a paddlewheel to grind some grain, we open a valve and water flows through a pipe. The size of the pipe nozzle determines the resistance to the flow, or exactly how much water comes out. The bigger the nozzle, up to the pipe's diameter, the more flow. The amount of flow is then shown on a flowmeter.

On a boat, the battery behaves very much the way the reservoir does, storing energy until we need it. When we do, we turn on a switch to start the flow of electricity. The amount of electricity that flows down the wire depends on the *resistance* of the appliance drawing the electricity. The less the resistance, the more electrical flow. A solid-state radio has a large resistance, and so it allows only a small flow. A 12V refrigerator has a small resistance, so it allows a large flow. This inverse relationship is a little tricky to remember.

### Electrical/Sailing Analogy

A second analogy compares electricity to sailboats. The analogy is fairly loose but the idea comes through nicely.

The mechanical energy of the wind is equivalent to the electrical energy from a battery. When we want to put this wind energy to work we turn on a switch, figuratively speaking, by raising the sails. The boat moves along through the water as electricity moves through a wire. If there was no water resistance on the hull we would go faster and faster. But in reality we have lots of resistance (often more than we should) that governs the boat's speed through the water and limits the useful mechanical work of transporting us from one place to another. In the same way the resistance in an electrical circuit determines the flow of electricity.

## CONCEPTS

Let's look at five basic electrical concepts.

### Voltage

The force or pressure that moves electrons along a wire, measured in volts with a voltmeter. The greater the voltage the more force there is. It's similar to the water pressure in your house.

### Current

The flow of electrons along a wire, measured in amperes (amps) with an ammeter. The greater the amperage the greater the flow of electrons. Similar to the amount of water flowing through the pipes in your house.

### Resistance

The opposition to the flow of electrons, measured in ohms with an ohmmeter. The more resistance, the lower the current. Similar to the size faucet at your sink—or the number of barnacles on your hull.

### Power

The amount of energy, or the rate of work done by electricity, per unit of time

$\frac{\text{ENERGY}}{\text{TIME}}$, measured in watts. Appliances are rated in watts, that is, the power they consume. For example, a 12V fluorescent cabin light may consume 12W of power while a 12V refrigerator may use 60W or more.

## Energy

The basic concern of this book. Represents how much power has been produced or consumed in a given amount of time, measured in watt-hours with a watt-hour meter (the electric meter outside your house or at a dock). If the cabin light mentioned earlier is left on for one hour, it will consume 12 watt-hours of energy, just as the 12V refrigerator would consume 60 watt-hours. As an interesting comparison, a typical American home uses 750,000 watt-hours of electricity each month!

The formulas below are called Ohm's Law after the German scientist George Ohm, who in 1827 discovered the relationships between electrical current, resistance and voltage. Ohm's Law shows how you can discover a value for either Current (I), Resistance (R) or Voltage (V), if you know the other two values. An easy way to remember this infinitely useful formula is to remember the triangle $\frac{V}{IR}$. Cover up the symbol for which you are solving and perform the mathematics indicated. Cover V, leaving I and R next to each other: multiply. Cover I, leaving V above R: divide. Cover R and V is above I: divide. Or, just remember the quickie formulas below. They'll prove invaluable when you're troubleshooting, trying to figure out what size wires to use, or noodling the effects of raising the wattage of your cabin lights.

$$V = IR \text{ or } I = \frac{V}{R} \text{ or } R = \frac{V}{I}$$

The next series of formulas explain a relationship of great utility to sailors. It defines the *charging rate* (power) of the generator and the *consumption rate* (power) of the appliances. If we multiply Current (I) by Voltage (V) we get a value for Power (P). For example, a generator producing 10 amps (I) at 16V (V) has a *power output* of 160W (P). Similarly, if a 12W (P) light bulb is getting power from a 12V (V) battery it will "draw" a current of 1 amp (I).

$$P = VI \text{ or } \frac{P}{V} = I \text{ or } \frac{P}{I} = V$$

or, if we substitute IR for V, then $P = I^2R$.

## INSTRUMENTS

Now let's look at some simple, inexpensive tools that help make electricity more visible.

### Test Light

A test light, or continuity tester, is the simplest and least expensive tool for testing electrical circuits. You can easily make one yourself or buy one for about $2. It is simply a 12V bulb with wires soldered to it (or in a socket, if you have one) with alligator clips at

*Laying the Groundwork*

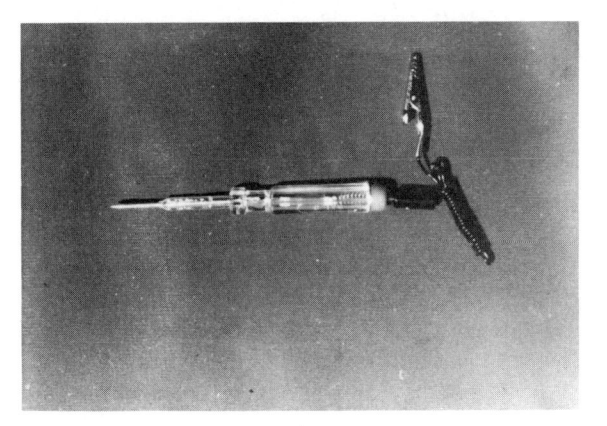

**Fig. 2-B.** A low-cost test light with detachable cord and alligator clip.

the end of the wires. A test light tells you if there is continuity in an electrical circuit, meaning no breaks in the wire or loose contacts. A whole circuit can be tested by putting one wire of the tester on the negative pole of the battery (or on the wire leading from the negative pole) and checking to see if current flows at various points in the circuit

**Fig. 2-C.** A typical test light application.

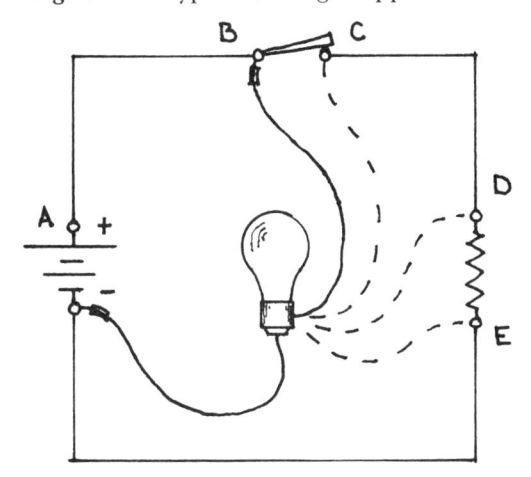

by touching each point with the other lead. In Fig. 2-C, start your test at B. If the bulb lights up, the wire from A to B is fine. Next, place it on C. If the bulb lights up with the switch closed, the switch is OK., etc.

The brightness of the bulb on the test light also will roughly indicate a battery's state of charge or the resistance in the circuit under test. To get exact readings you will need a voltmeter or hydrometer.

## Voltmeter

A voltmeter is used to read the voltage output from your generator or alternator or battery. It comes in various scales of voltage of which the best for a boat is a 0—25 DC scale. You can mount it near your electrical panel or get a hand–held model. Some electrical panels already come with a battery-condition indicator reading in volts. To use the voltmeter in a circuit, place it across a load, as shown in Fig. 2-D, making sure the leads are hooked

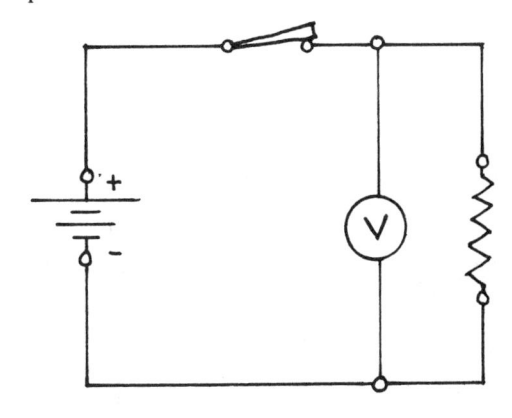

**Fig. 2-D.** Schematic diagram of voltmeter in a simple circuit.

*Basic Electricity*

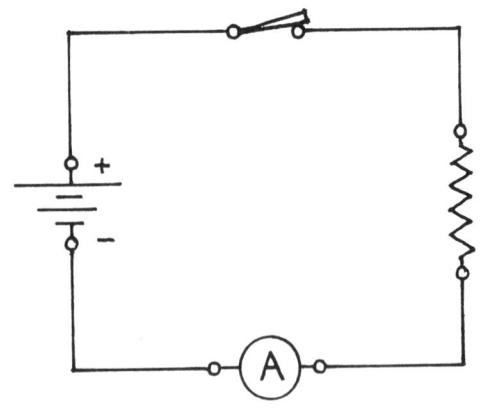

**Fig. 2-E.** Schematic diagram of ammeter in a simple circuit.

correctly, *ie.* positive to positive (red to red), negative to negative (black to black). Cost is about $15.

### Ammeter

An ammeter reads the current flow in any circuit, or the total draw, or load, of several things operating at once—or the current produced by a generator or alternator. It, too, is a small meter available in various scales, of which the best for a boat is 0–5, 0–10, or 0–20 DC scale, depending on the amount of current you are measuring. Like the voltmeter, it

**Fig. 2-F.** Multimeter with positive (red) and negative (black) test leads.

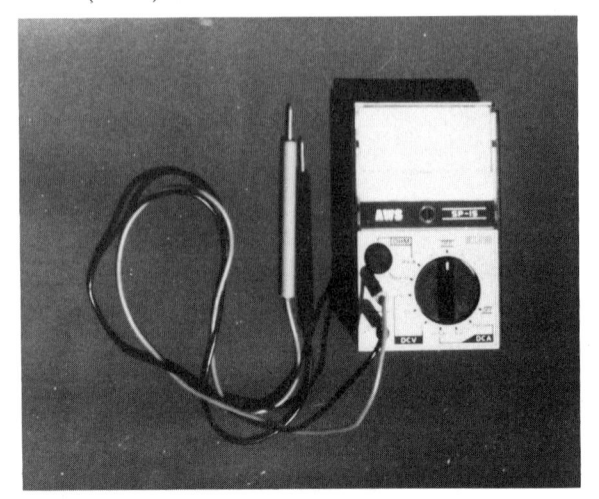

can be easily mounted where convenient or kept for hand-held use. To use, an ammeter is connected in series with the load in a circuit, as shown in Fig. 2-E, and located anywhere along the wire. Polarity must be observed with an ammeter. Cost is about $15. Note: We've seen ammeters that are included in an instrument panel that had a 0–100 amp scale! A bit of nonsense for a sailboat that uses 10–15 amps maximum—unless your boat is fitted with heavy electrical gear, such as a windlass or electric winch. Our wind and water unit came with a control box that has voltmeter and ammeter mounted side by side for convenient monitoring of system performance (See Fig. 10-D).

### Ohmmeter

An ohmmeter usually is combined with a voltmeter and ammeter in a hand-held instrument called a multimeter. The internal battery of an ohmmeter produces a small current that enables you to measure the resistance of the wiring or individual appliances. Readings will be in ohms. *Always* shut off power to the circuit, *i.e.* disconnect the battery, before using the ohmmeter. To use, place the two leads of the meter on the two terminals of the appliance or two ends of the wiring in the circuit. The resistance of individual appliances can be measured, but we most often use our ohmmeter for testing the continuity of circuits that are not connected to an external voltage source.

### Multimeter

A multimeter is a hand-held instrument with a selector switch used to read either voltage, resistance or small amounts of current. We have a standard needle readout model that we bought for about $12. A digital model, which is much more accurate, will cost $50 or more.

## TYPES OF ELECTRICITY

There are two main types of electricity, direct current and alternating current. A boat's electrical system is direct current (DC), while a hook-up outlet at a dock or the electricity in your house is alternating current (AC).

### Direct Current

DC means the electricity or electrons in a wire always move in one direction. Benjamin Franklin gave us the names "positive" and "negative" for terminals of a battery, and assumed current moved from positive pole to negative pole. Since then, convention shows current moving in this direction. However, electrons actually are flowing the other way. But don't be too concerned with direction of flow. What is important is that all DC appliances and electronics (except incandescent lights) are marked with positive and negative leads and *must* be connected correctly. An electric motor (such as a bilge pump) will run backwards if the leads are reversed, and electronic gear will either blow an internal fuse or become damaged.

An important fact to remember is *only direct current can be used to charge a battery*. Although boat batteries are rated at 12V, the voltage of a particular battery will vary slightly depending on the state of charge.

### Alternating Current

AC means the electron flow, or electrical current, moves first in one direction, then in the opposite direction, "alternating" in this manner at 60 Hertz (H) frequency, or 60 cycles every second. In your house, voltage is about 110V AC. Electric companies prefer alternating current at high voltage because it is more efficient to produce and transmit electricity this way. Alternating current must be "rectified", or turned into direct current through the use of diodes, before it can be used to charge a battery.

**Fig. 2-G.** A few of the more common schematic symbols used for boat circuits.

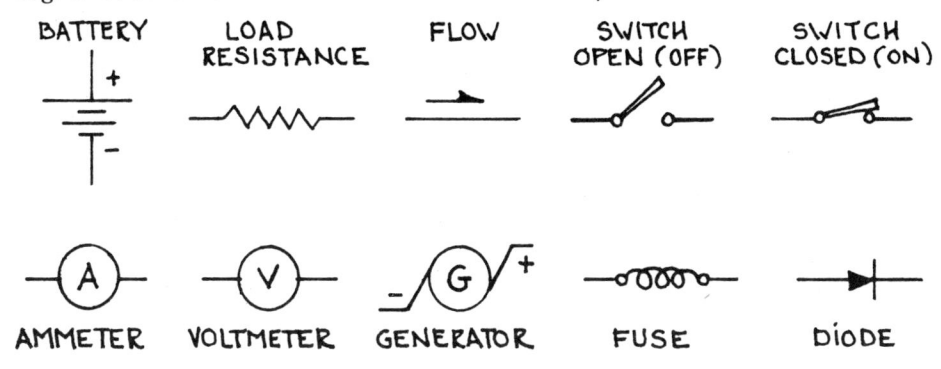

## A SIMPLE 12V CIRCUIT

An electrical circuit is formed by connecting a voltage source (battery) to a load of some sort (appliance) by a conducting wire (conductor), usually copper. A circuit is greatly simplified by putting it in schematic form using the symbols below. If you want to understand or find the trouble spot in a given circuit, draw it out in a schematic first, and the problem will probably become clear. See Fig. 2–G.

In schematic form, a simple circuit from a boat battery to one instrument panel function (e.g. a compass light) looks something like this:

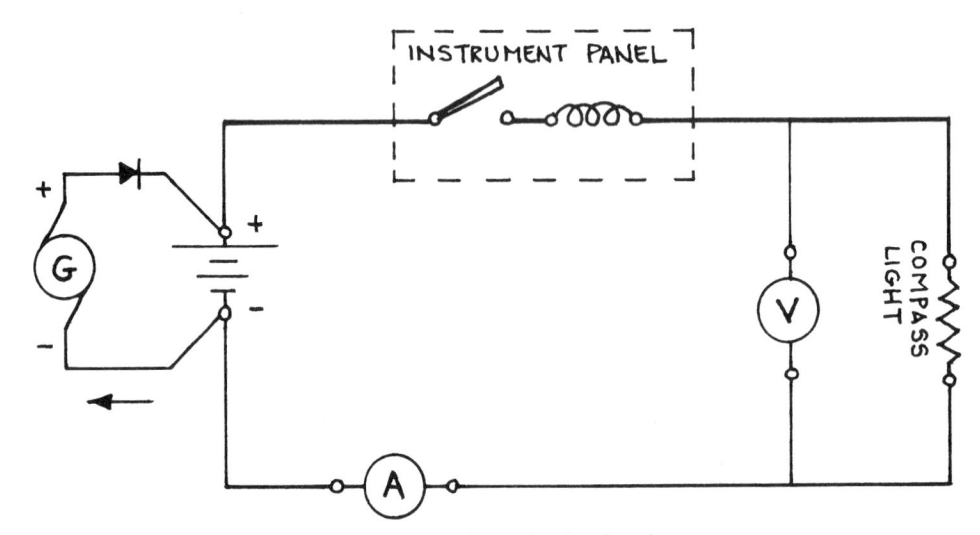

**Fig. 2–H.** A simple schematic diagram for a single circuit.

(We're staying with convention here and showing the current flow from positive to negative pole.) Note that some of the electron flow from the generator will pass through the circuit if there is a load, while the excess will flow to the battery for later use.

## BATTERIES

Of the many types of batteries available, the one we are concerned with on boats is the lead-acid type most often rated at 12V. These batteries have 6 2V cells, which are connected in series to provide 12V. The cells all contain the following items:

1. Positive plates made of a lead grid carrying dioxide or lead peroxide.
2. Negative plates of porous spongy lead.
3. A dilute sulfuric acid solution surrounding the plates called the *electrolyte*.

### Characteristics

The characteristics of the 12V battery are:

1. The terminal voltage actually varies, depending on state of charge, from 11V

14
*Laying the Groundwork*

near discharge to approximately 13.8V at full charge with no load. A voltmeter will tell you the voltage of your battery.

2. The voltage of the battery is directly dependent on the strength of the sulphuric acid solution, or electrolyte. A measure of the strength of the electrolyte is its specific gravity, which varies from 1 at complete discharge to 1.26–1.3 at 80°F fully charged. The rating of the electrolyte's specific gravity at full charge varies slightly with each type of battery. We have a large marine DieHard that is at 100% of charge when the specific gravity is 1.26. Below is a chart similar to one that came with our battery explaining the relationship between specific gravity and state of charge. Some batteries will not reach full charge until the specific gravity has reached 1.3. The difference is strictly in the type of electrolyte and does not affect the capacity. The specific gravity is measured with a hydrometer.

| State of Charge | Specific Gravity | Hours to Recharge @ 10A | Hydrometer Balls Floating |
|---|---|---|---|
| 100% | 1.26 | – | 4 |
| 75 | 1.235 | 3 | 3 |
| 50 | 1.210 | 7 | 2 |
| 25 | 1.185 | 10 | 1 |
| 0 | 1.16 | 24 | 0 |

3. When measuring battery charge, take into consideration whether you are generating electricity and charging the battery, or using electricity and discharging it. Readings will be higher when the battery is being charged and lower when under load or being discharged. The most accurate measurements are made a short time after charging has stopped and no current is being drawn. When charging with a low output generator (less than 10A) the electrolyte will remain stratified until the battery is near full charge. As a result, hydrometer readings will be lower than actual until gas bubbles have a chance to mix the electrolyte.

4. The specific gravity should not vary much from cell to cell. If one cell is *much* lower than the others, it probably is dead. With the bulb-type hydrometer this will show up as no balls floating. With a specific gravity readout model, the number on the float will show much lower than the other cells. A dead cell will lower the total voltage, and the battery should be tested at a shop and replaced if necessary. One dead cell often disables the entire battery by preventing a flow of electricity through the battery.

5. A battery's capacity is rated in amp-hours, or how many amps you can draw from it in a given time. A typical battery has 100 amp-hours (amps X hours) from which, theoretically, you can draw 10 amps for 10 hours, or 5 amps for 20 hours, or 1 amp for 100 hours. (Some authorities use the term "ampacity". It means the capacity of the battery in amperes. One such authority is the American Boat & Yacht Council, whose standards are widely used.)

6. You will get *more usable amp-hours* from a

*Basic Electricity*

battery if you draw the currrent slowly than if you take it out quickly, as illustrated by the chart below.

| Amperage Draw | Hours of Usable Power | Total Amp Hours |
|---|---|---|
| 2.5 | 36.0 | 90.0 |
| 5.0 | 14.5 | 72.5 |
| 10.0 | 6.3 | 63.0 |
| 15.0 | 3.8 | 57.0 |
| 20.0 | 2.7 | 54.0 |
| 25.0 | 2.1 | 52.5 |

7. A true marine battery is known as the "deep-cycle" battery. It's the best kind for sailboats. Unlike your car battery, which gets charged every time you turn the engine on, and therefore is not deep cycled, your boat battery may go many hours between chargings, regardless of the type of charging system you use. A deep-cycle battery will have thicker plates and more dense active material than shallow-cycle types.
8. To be specific, then, although a battery is rated at 100 amp-hours, you get the

**Fig. 2-I.** The chemical changes inside a lead-acid cell (battery):

a.) The sulphuric acid ($H_2SO_4$) in the electrolyte is "ionized" during discharge:

$$H_2SO_4 \underset{\text{Charge}}{\overset{\text{Discharge}}{\rightleftarrows}} 2H + SO_4$$
sulphuric acid → hydrogen ions + sulphate ion

b.) The chemical reaction at the negative electrode is:

$$Pb + SO_4 \underset{\text{Charge}}{\overset{\text{Discharge}}{\rightleftarrows}} PbSO_4 + 2\epsilon$$
lead + sulphate ion → lead sulphate + electrons from hydrogen ion

c.) The chemical reaction at the positive electrode is:

$$PbO_2 + 4H^+ + SO_4 + 2\epsilon \underset{\text{Charge}}{\overset{\text{Discharge}}{\rightleftarrows}} PbSO_4 + 2H_2O$$
lead dioxide + hydrogen ions + sulphate ions + electrons → lead sulphate + water

*Laying the Groundwork*

longest service if you use only the top half, or 50 amp-hours, of its capacity before charging it up again. Any battery, including a deep-cycle one, will give out or lose much of its rated capacity if it is repeatedly taken down to near complete discharge. For an interesting exception to this rule, see Chapter 10, Batteries.

Let's see what chemical changes take place inside a battery as it is charged and discharged. See Fig. 2–I.

As you discharge the battery, the concentration of sulfuric acid *decreases,* the specific gravity *decreases,* and your voltage drops. Also, hydrogen ions are produced. A form of charged hydrogen particle, they are harmless in themselves, but they quickly combine to form ordinary hydrogen gas which, when mixed with air, is a powerful explosive. That's why you keep flames and sparks away when a battery is being charged, and why batteries must always be kept in a well-ventilated space. Under ordinary charging conditions on a boat, never take the caps off the cells. This will minimize the access of air to the hydrogen-laden spaces above the plates in the cells. Battery caps do have vents, which ought to be kept clear, to permit the escape of gases.

## Hydrometers

A hydrometer is a very simple, inexpensive way to measure the specific gravity of each cell in your battery. While some maintenance-free batteries are permanently sealed

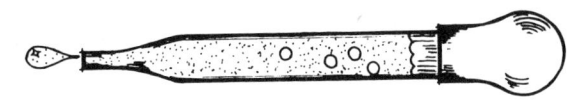

**Fig. 2–J.** A "floating balls" type hydrometer.

(no vent caps) so you can't take readings, most have cells with removable caps. The least expensive hydrometer has four balls inside a clear tube. To use, squeeze the bulb and dip the tube nozzle into the electrolyte in a cell. Don't splash sulfuric acid!. Take some of the solution into the tube. The number of balls that float indicates the state of charge—the more floating the greater the charge. A more expensive model follows the same procedure but gives direct readings of specific gravity. See Fig. 2–J.

## Multiple Batteries

Many boats have more than one battery on board. Two batteries can be connected together in parallel, as shown in Fig. 2–K. Together they will act as one big battery while the voltage remains the same.

The generator or alternator leads may be connected to the terminals of either battery or at any point along the main wires (as shown by the dotted lines). In either case, the generator will charge the batteries equally, and the batteries will tend to even out the possible variations in voltage from a generator.

More commonly, the batteries are isolated through a main switch so you can use either one or both, often reserving one for starting the engine and the other for appliances. See Fig. 2–L.

17
*Basic Electricity*

**Fig. 2-K.** Schematic of batteries connected in parallel.

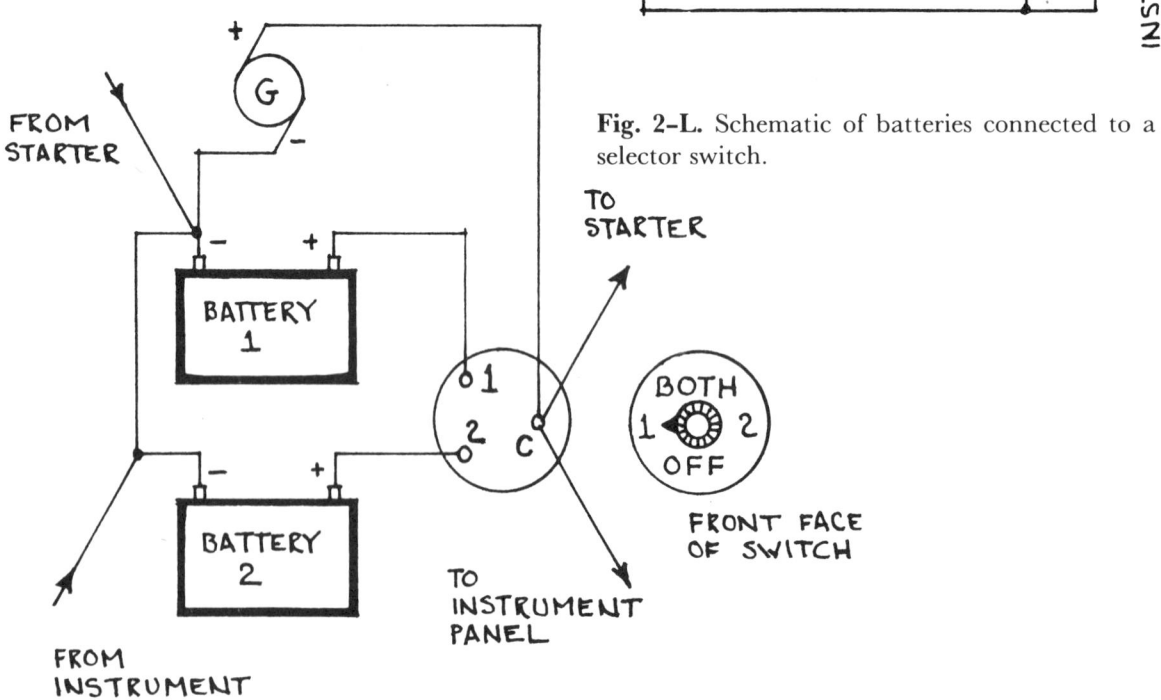

**Fig. 2-L.** Schematic of batteries connected to a selector switch.

If you connect two or more batteries this way you must be careful of several things:

1. The main switch must *not* be in the "off" position while generating. This will damage the alternator or prevent current from reaching the batteries if using an alternative method of charging.
2. If one battery is discharged, or at a low state of charge, it will affect the other battery(s) when the main switch is turned to "both". The batteries are at that time temporarily connected in parallel and the discharged battery will draw current until the batteries are at equilibrium (same state of charge). In Fig. 2–M columns of water separated by valves are a good illustration of this concept.) When the valves are opened and the water tubes interconnected, the water will seek the same level in all three columns.

Current will act in the same way as it attempts to equalize the state of charge under the influence of electrical pressure, or voltage.

These difficulties warrant a search for something better. Several products on the market eliminate these problems while still allowing the batteries to be isolated. One product, the Sea-Isolator by RGM Industries, Inc. offers two models: The BI 2-60 (about $26) allows two batteries to be charged equally while keeping each battery circuit isolated with respect to battery drain. See Fig. 2–N. The BI 3-60 (about $29) does the same for these batteries. Similar products are offered by the Guest Corporation (the Splitter) and by Spa Creek Instruments (Charging Diodes).

If you are handy with a soldering iron, you can isolate the battery circuits yourself by using diodes. This procedure, along with other do-it-yourself projects for the sailor, is outlined in "The 12 Volt Doctor's Practical Handbook" by Spa Creek Instruments.

When a third battery is desired it may be connected to the others according to your needs. Three batteries may be isolated by

**Fig. 2–M.** Batteries seeking electrical equilibrium.

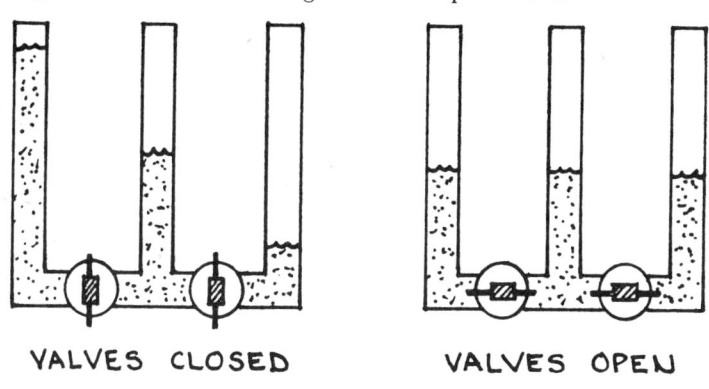

VALVES CLOSED    VALVES OPEN

Basic Electricity

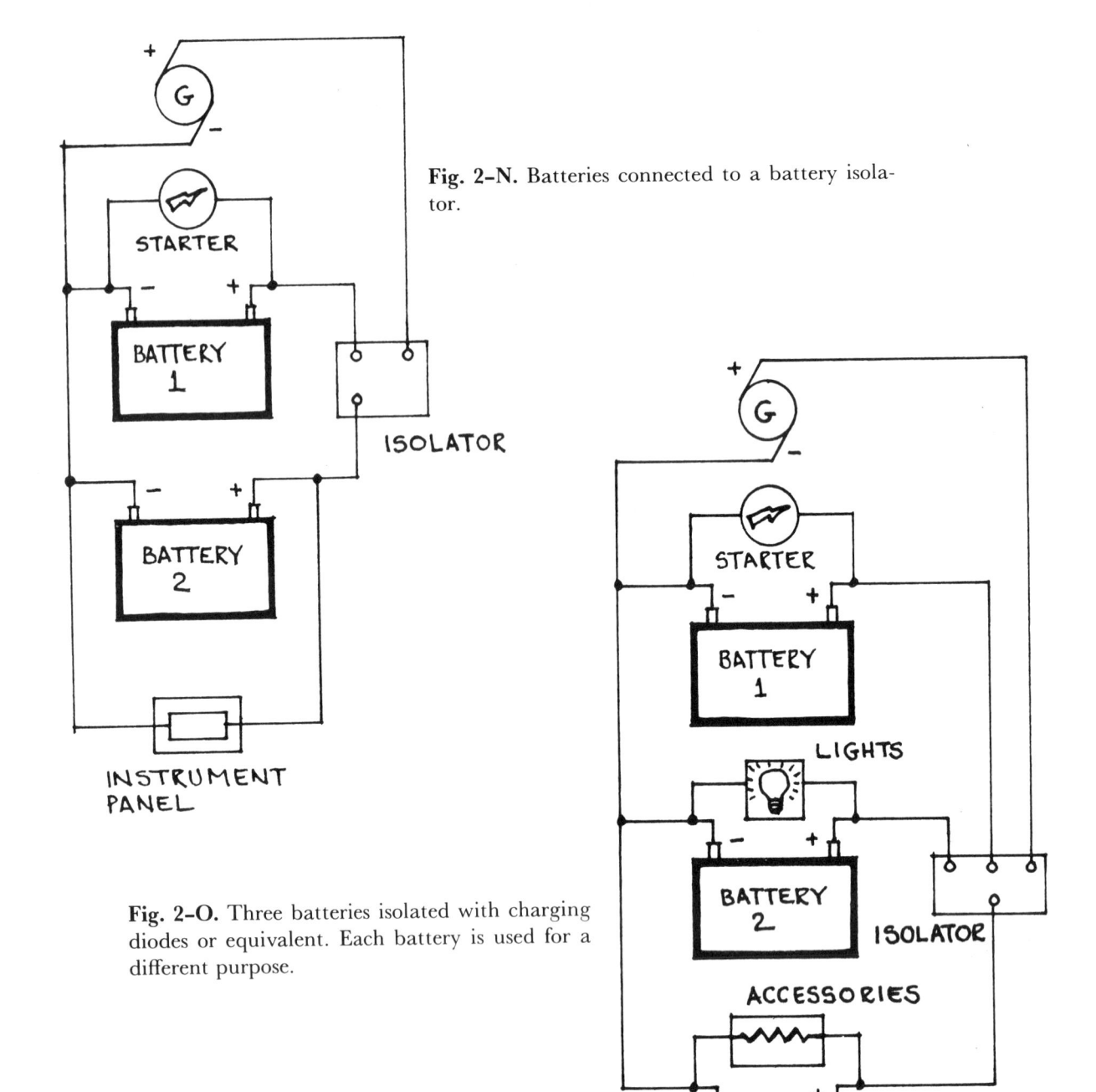

**Fig. 2-N.** Batteries connected to a battery isolator.

**Fig. 2-O.** Three batteries isolated with charging diodes or equivalent. Each battery is used for a different purpose.

20
*Laying the Groundwork*

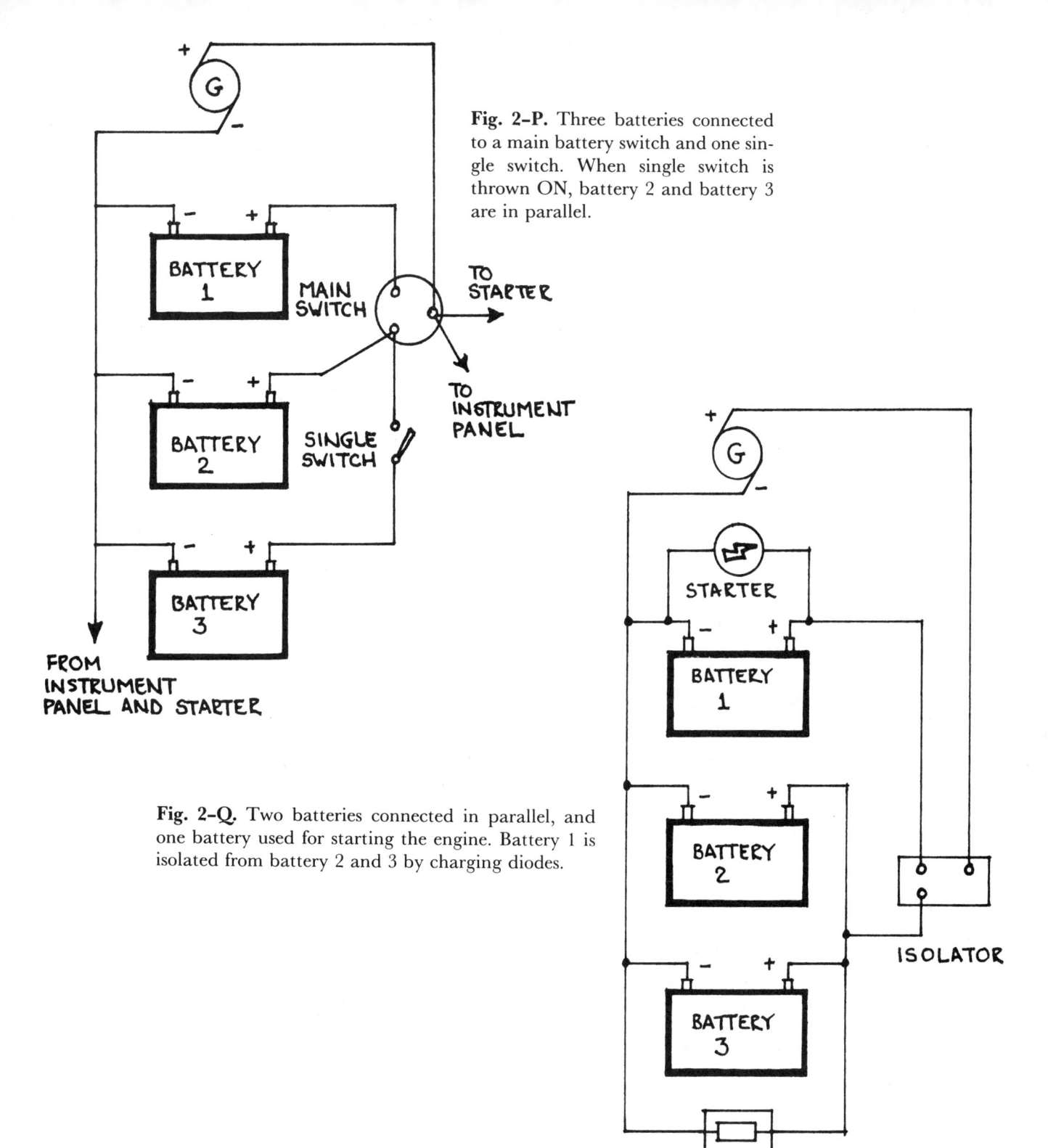

**Fig. 2-P.** Three batteries connected to a main battery switch and one single switch. When single switch is thrown ON, battery 2 and battery 3 are in parallel.

**Fig. 2-Q.** Two batteries connected in parallel, and one battery used for starting the engine. Battery 1 is isolated from battery 2 and 3 by charging diodes.

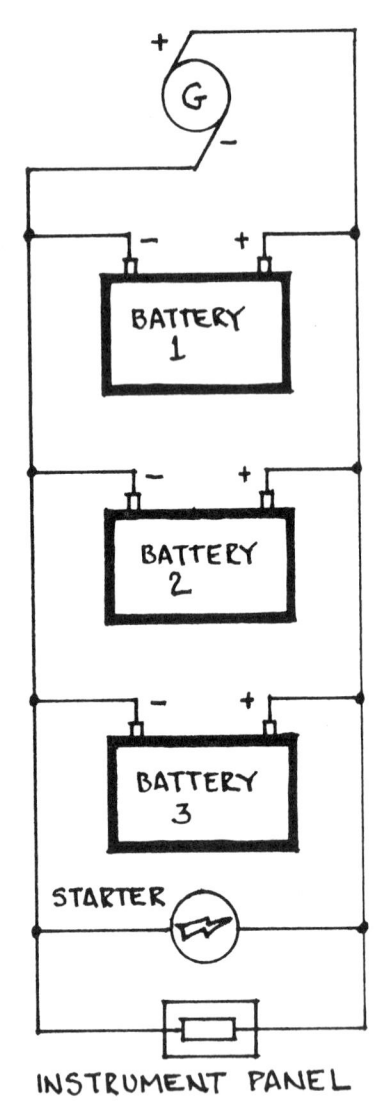

**Fig. 2-R.** Three batteries connected in parallel.

using a store-bought isolator or charging diodes (see Fig. 2-O) or a main switch with a separate toggle switch (see Fig. 2-P). Two batteries may be joined in parallel for appliances, leaving one available to start the engine (see Fig. 2-Q). Or all may be connected in parallel to act as one large battery (see Fig. 2-R).

## A BOAT'S ELECTRICAL SCHEMATIC

We are now ready to look at a typical boat's electrical schematic. Let's assume we have:

1. Two batteries connected through an isolating device.
2. An engine that requires a heavy current to the starter.
3. An alternator that supplies a charging current to the batteries.
4. An instrument panel supplying electricity to six functions, with a switch and a fuse for each.
5. An electric bilge pump that will operate independently of your main battery switch (for those times when the boat is left unattended).
6. An alternate energy generator added to the system.

## SAFETY

Many people are reluctant to tackle their boat's or home's electrical system because of the danger inherent in working with electricity. If you remember a few simple rules you can work with your boat's electrical system in complete safety.

1. Always keep in mind which pole of the battery is positive and which negative. The positive pole usually has a red wire

22
*Laying the Groundwork*

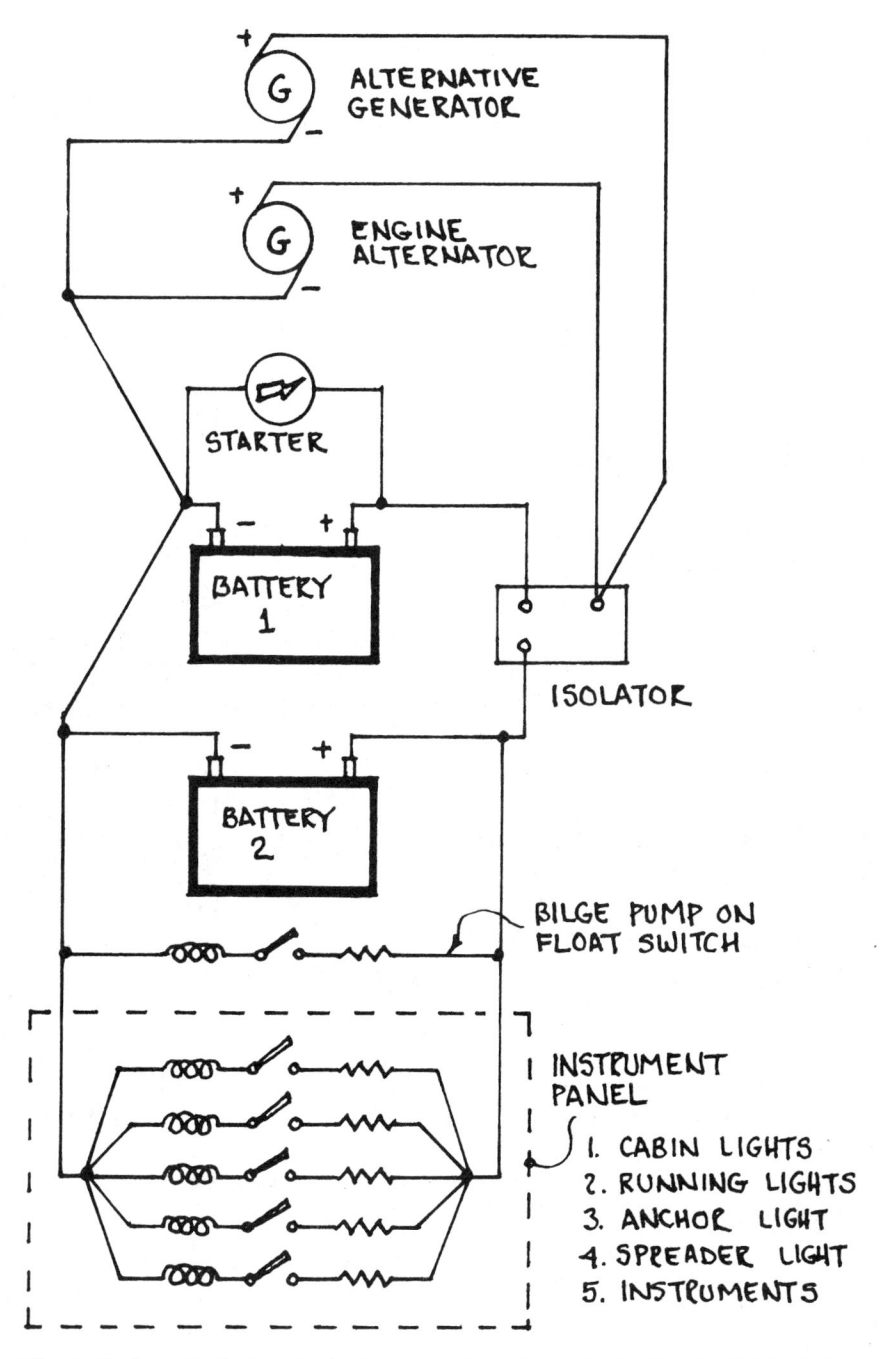

**Fig. 2-S.** A typical electrical schematic for a boat, including individual circuits on an instrument panel and an alternative generator.

attached and a "+" or "Pos" written next to it. The negative wire is usually black, and has a "−" or "Neg" written next to it. Both cables are heavy-gauge.

2. The electrolyte in batteries is *sulfuric acid* and should not come in contact with your eyes, skin, clothing, cockpit cushions, etc. We burned a perfect round hole through a pair of white pants the first time we handled our boat battery. If physical contact is made with the electrolyte, flush immediately with water, apply a solution of baking soda mixed with water, and cover with a dry gauze bandage. Do *not* apply petroleum jelly. For an acid burn to the eye, splash with water for five minutes (several changes of water in an eye cup are even better), apply two drops of castor oil to the eye, and cover with compresses dipped in a solution of one teaspoon of salt in a glass of fresh water (not seawater). Get medical help!

3. When a battery is being charged it gives off hydrogen gas, which is explosive. So don't take the battery caps off during charging, don't let sparks or a flame get near the battery and make sure the battery locker is well-ventilated. Make sure the cap vents are clear.

4. Never allow current to flow from the positive terminal of a battery directly to the negative terminal, e.g. by touching positive and negative wires from a battery together, or by letting a metal tool come in contact with both terminals at the same time.

5. The voltage on a boat (12V DC) is much lower than the voltage in a house (110V AC). You may touch one terminal of the battery at a time without fear of getting a shock. Your body is not a particularly good conductor but can become one in certain circumstances, e.g. when the skin is wet. Make sure you are not standing in water. It's also a good idea to insulate your hand with gloves and use tools with rubber-coated handles.

6. The main switch should be turned off before attempting repair of electrical circuits.

7. Never connect an ammeter directly across battery terminals, as it has little resistance and the meter may blow out. Make sure the ammeter is rated for the range of current you want to measure. If you are unsure, go to a larger value.

8. Never hook 12V batteries in series, as this increases the voltage to the combined voltage of the (two batteries—24V) which you probably don't need on a boat, and which can severely damage 12V devices. You can, however, hook two 6V golf cart or truck batteries in series to geet 12V. A number of sailors have found 6V batteries to be less expensive and better performers.

You now have the ability to understand and maintain your boat's electrical system, just as you maintain her hull and sails. Although a complete understanding of electricity is not necessary, it increases the pleasure in operating an alternate energy system. You know how the energy is being produced, where it is going, how the battery stores it, and at what rate it is being used up. This knowledge will help you to maintain your system in peak condition and avoid or diagnose trouble spots. Most of all, it will give you an appreciation for the simplicity and continuity of alternate energy electrical generation.

# CHAPTER 3

# MARINE ALTERNATE ENERGY EQUIPMENT—A GENERIC DESCRIPTION

Many different kinds of alternate energy device are available today for boats. Each has its own particular design, philosophy, construction and mounting technique. Here we will familiarize you with the gear and the terms, to provide you with a good background for the Reference Section. There, you can learn the theory (Chapter 16) read about the products available on the U.S. market (Chapter 17), and evaluate the merits of each product (Chapter 18).

## SOLAR

Various ways exist to capture solar energy and put it to good use. The usual sequence of events is to first trap the sun's rays and convert this radiant energy into a more usable form. It then can be transported to a storage medium where it remains until we are ready to use it. The devices that trap and convert solar energy are called "solar panels". Some solar panels trap the sun's light energy (short wave radiation) by converting it to heat energy (long wave radiation) when it is absorbed by a dark surface. One or more layers of glass keep this heat from escaping until it can be taken to a storage area by a moving fluid (water, air, or other transfer fluid). The hot water "sun shower" is a simple version of this type of solar panel, combining collector and storage medium in one neat package.

In this book our focus is on another type of solar panel, the photovoltaic panel, which converts light energy directly into electricity. Instead of using a dark surface to absorb the sunlight, photovoltaic panels use a special material (usually single-crystal silicon) that has been chemically treated and formed into thin wafers called solar cells. Solar cells are connected in series to provide the voltage necessary to charge a battery. For our purposes we will henceforth refer to photovoltaic panels simply as solar panels.

## WIND

Wind generators are composed of two basic parts. The first is a propeller or equivalent device that transforms the mechanical energy of moving air into the mechanical energy of a rotating shaft. The second basic part, an electrical generator, transforms the shaft's rotary motion into electrical energy. The amount of electrical energy we can extract

## Classification of Solar Panels

BY OUTPUT:　　Solar panels are rated by the maximum (or peak) electrical power they produce. They are available from small trickle-charge panels that have a 7–10W rating, up to large, 66W modules. The 35W panel has become the standard size for land use and, generally speaking, is the least expensive in terms of dollars/watt. On today's marine market you can purchase 7, 9, 10, 20, 30, 35, 40, 44, and 66W panels.

BY MODEL:　　*Standard solar panels* are designed and primarily used for land-based applications. Since newer models are ruggedly built with a long life expectancy, they are also marketed for marine use. We have one of these on our boat, *Kjersti*. The general characteristics of models suitable for marine use are a smooth, tough, tempered glass covers with very high light transmission. An aluminum frame usually is at the perimeter, to prevent undue bending stress on the cells. They are not made to be walked on, although they can be in an emergency. They usually contain 36 round or half-round silicon solar cells. Some use rectangular "polycrystalline" cells that can be fitted closely together in solar modules. These polycrystalline cells can be made from a less pure silicon, but their dense configuration makes up for their lower efficiency. Each panel has 36 cells (occasionally more or less), at a 1/2V each, to total the required voltage to charge a battery even on overcast days when the current generated is lower.

*Marine solar panel* manufacturers alter standard solar panels to produce marine-grade equipment. These panels are made to be walked on. Some use a shatterproof textured Plexiglas or Lexan cover that provides fairly good footing when stepped on. Some use a smooth, very strong, virtually indestructible cover of Tedlar, or equivalent polymer material. Some marine panels also eliminate the frame to achieve a very thin panel, and also use stainless steel as a perimeter edging and backing plate for additional strength and resistance to flexing. Silicon cells themselves are strong but may fracture when exposed to repeated bending. (Silica is the same material used to make glass, which, as we know, is also strong but susceptible to fracture.) Finally, many manufacturers put a teak border around the

**Fig. 3-A.** Four Free Energy Inc. solar panels mounted on the cabin roof of a sloop.

**Fig. 3-B.** A Kyocera marine grade solar panel with teak trim (courtesy of Kyocera International, Inc.).

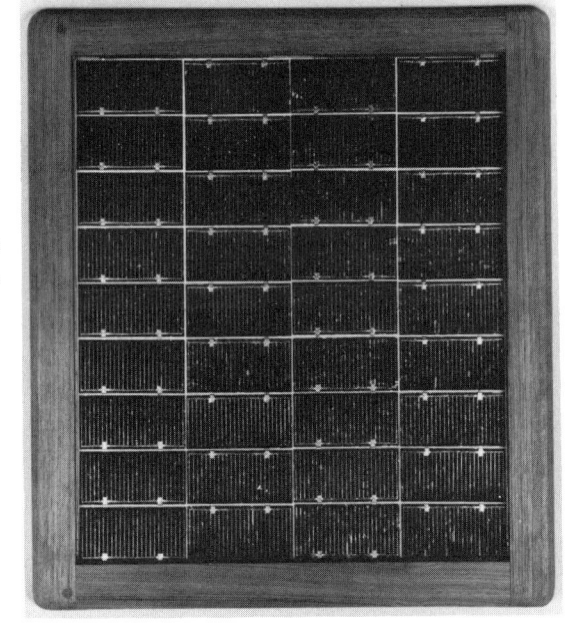

**Fig. 3–C.** A Solec Solarcharger solar panel with polymer cover (courtesy of Solec).

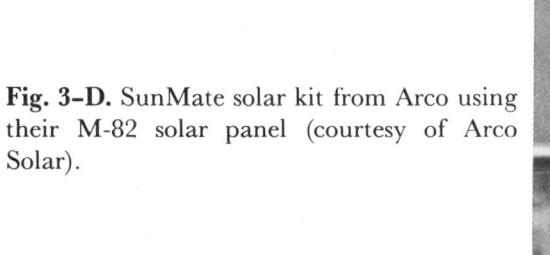

**Fig. 3–D.** SunMate solar kit from Arco using their M-82 solar panel (courtesy of Arco Solar).

**Fig. 3–E.** Marine grade solar panel from PDC Labs mounted on the stern rail (courtesy of PDC Labs).

**Fig. 3-F.** Our solar panel gathering energy under sail in the Bahamas.

**Fig. 3-G.** Set on the Bimini, our solar panel is out of the way and has good exposure to the sun.

panel to help it blend with the boat's trim. While in no way affecting the panel's performance the trim does provide an elegant feature.

BY MOUNTING TYPE:

*Deck Mount*—A panel mounted flush (or as close as possible) on the cabin top or deck, leaving room for air circulation behind the panel.

*Hatch Mount*—A small panel mounted on top of an opaque hatch cover (not one that lets light below!). This locates the panel in an area of the boat that does not have heavy foot traffic and that usually is well exposed to the sun.

*Stern Rail Mount*—A panel laid horizontally on the stern rail and fastened with U-straps or equivalent. If fastened correctly it can be pivoted inboard or out to follow the sun. A gimballed frame achieves the same results. Stern rail mounting kits are available from several manufacturers.

*Moveable Mount*—This is a very good option. This mount permits the panel to be adjusted anywhere on the boat so it always faces the sun. A semi-permanent mount also is an option, allowing you to remove and direct the panel towards the sun when at anchor and to lock it in place under sail.

*Laying the Groundwork*

from the wind depends on the following things:

*Wind speed*—The power output of a wind generator is proportional to the cube of the wind speed. A 2-knot increase in wind speed will provide an eight-fold increase in power output.

*Propeller size*—The power output of a wind generator is proportional to the square of the diameter of the propeller.

*Propeller efficiency*—This is a measure of how good the propeller is at transforming wind energy into the rotary motion of a shaft. The efficiency varies dramatically with different types of propellers.

*Generator efficiency*—This is a measure of how good the generator is at transforming the mechanical energy of a rotating shaft into electrical energy. The efficiency will vary with wind speed and the type of generator used.

## Maximum Theoretical Power

Even in ideal conditions, only about 59% of the power available in moving air can be converted into the rotary motion of a shaft (as shown in Chapter 16). In practice, the efficiency of the propeller and generator further reduce the usable power, typically to 25-30% of the available power in the wind before it came in contact with the propeller.

## Density of The Fluid

For sailors, this is the density of air at sea level.

---

### Classification of Wind Generators

BY ENERGY TRANSFORMATION METHOD:

*Propeller-type wind units* include all of the wind generators now on the marine market. The American wind units use mostly 2-bladed propellers similar to ones used on airplanes. Several 3-bladed versions are available that use a different type of airfoil. See Fig. 3-H. The two British units marketed in the U.S., Ampair and LVM, are multi-bladed propellers reminiscent of farm water-pumping windmills. These multi-bladed propellers are less efficient, but do not require special overspeed protection in high winds. All these units transfer horizontal wind energy into rotary motion about a *horizontal* axis.

The former Eodyne wind unit and the mini Windtrap, neither of which are any longer in production, used a different wind principle. They transferred horizontal wind energy into rotary motion about a

*Marine Alternate Energy Equipment—A Generic Description*

**Fig. 3–H.** A Fourwinds 3-bladed wind generator mounted on a pole (courtesy of Everfair Enterprises).

*vertical* axis with a Savonius-type wind generator, which functions much like an annemometer. The power potential is fairly low in this type of unit.

BY PROPELLER SIZE AND CONFIGURATION:

*Large-propeller wind units* are any unit with a 3′ diameter propeller or larger. These include all of today's American marine wind generators. As the generator on most large-prop wind units can be damaged by excessive speed, these machines must be manually or automatically shut down in strong winds (25-30 knots). Methods of automatically governing the propeller speed include spring-loaded deflection blades, feathering propellers, wind-deflecting tail vanes, mechanical clutches, and spring loaded tilt-up generator and propeller assembly.

**Fig. 3–I.** Hamilton Ferris offers an optional "air brake" as an overspeed governing device for his wind generators.

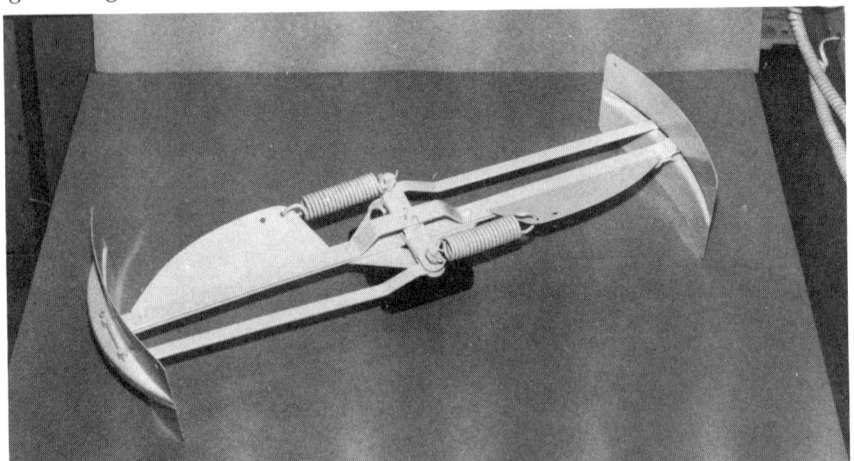

*Laying the Groundwork*

**Fig. 3-J.** The thermax wind generator, suitable for a boat or land use (courtesy of Thermax).

*Small-propeller wind units* are those with propellers smaller than 3′ in diameter, often British-made units that are multi-bladed. These units can withstand very high winds without producing speeds harmful to the generating unit. They sacrifice higher outputs for safety and the ability to be completely and inherently self-tending.

BY DRIVE SYSTEM:

*Direct drive wind units*—With one exception, all of today's marine wind units are *direct drive,* with the propeller coupled directly to the generator shaft. This avoids the need for intermediate gears or drive systems. Simplicity of operation and maintenance and reduced production costs account for the popularity of these units. Since the speed of the propeller is also the speed of the generator shaft, special generators must be used that operate at relatively low speeds. (See Chapter 16.)

*Geared drive wind units*—To our knowledge, there exists only one geared-drive marine wind system, the Silentpower model from

**Fig. 3-K.** A reconditioned farm wind generator mounted to the mizzen mast of *Different Drummer*.

Wind Turbine Industries. Its geared drive system permits use of a standard high-output marine alternator as its generating source. A few older geared units are still around. We saw one in Royal Harbor, Bahamas, that was a reconditioned Wincharger originally used on a farm for generating electricity. It was mounted to a mizzen mast and had a 10′ diameter propeller with a 9-to-1 gear ratio!

The Web Charger (now out of production) was another marine wind unit not using direct drive. We owned one before our present Hamilton Ferris unit. The Web Charger had a 5′ diameter propeller rotating on its own journal bearing, with rotary motion transferred to the generator by a small car fan belt with a 2-to-1 ratio. For every rotation of the propeller, the generator rotated twice. In theory, this is very sound reasoning, for all the considerable lateral forces of the large propeller were taken by the strong journal bearing, leaving the generator bearing nicely isolated. Potentially, the generator also could get the necessary speed with half the rotary speed of the propeller, resulting in a quieter machine with higher output. In practice, we found the friction of the belt kept the propeller from operating in anything less than a steady 10-12 knots. The output of our particular generator also was less than the manufacturer's rated output.

BY MOUNTING TYPE:

*Rigging-suspended mounts* allow the wind generator to hang suspended in the foretriangle between the forestay on a sloop (inner stay on a cutter) and the mast, or in the main triangle. It is raised in place with a halyard and tied off at the bottom with three guy wires to

**Fig. 3–L.** Our Hamilton Ferris wind generator suspended in the foretriangle.

prevent swinging. The unit is attached with swivels at the top and bottom so it can rotate 360°. Where there isn't enough room to rotate the full circle, the unit can be tied off, but always it should be able to rotate 90° to the wind for effective shutdown. At anchor, this shouldn't present any problem as long as the boat swings into the wind. The mount must be removed when underway, although the foretriangle mount could be tied off and left in place when under motor or using the mainsail alone, and provided the seas were very

**Fig. 3–M.** An Aquair 50 wind generator operating at anchor.

*Marine Alternate Energy Equipment—A Generic Description*

**Fig. 3–N.** Home-built wind generator fastened to the inner shrouds. This type of mount relies on the boat (on anchor or mooring) to track the wind.

smooth. Any bounding around can easily damage the unit and create a hazard for people on board.

*Permanent or fixed mounts* are wind units that are permanently mounted in a fixed bracket, usually on the forward face of the mizzen mast. (See Fig. 5–I.) We also have seen fixed wind units

**Fig. 3–O.** A propeller and generator kit fix-mounted on the bow pulpit.

**Fig. 3–P.** Ampair multi-bladed wind generator mounted on a pole.

mounted on the bow pulpit and across the upper part of a pair of inner shrouds. (See Figs. 3–N, 3–O.) Since they are face forward and do not rotate, most fixed, large-prop generators rely on the boat itself to track the wind at anchor or on a mooring. Several custom mounts we've seen allow the wind unit to rotate 30° to each side, similar to a rotating fan. This type is much more responsive to tracking the wind, especially if the boat tends to sail at anchor.

Because of their small size and self-tending ability, the British units often are mounted on a short pole at the masthead, or off the upper part of the mast similar to a radar mount. (See Fig. 3–P.) This allows them to rotate 360° and track the wind anytime, even while anchored in a current or sailing off the wind (a truly fixed mount is almost useless on a downwind sail). Any wind unit that attempts to track the wind must have a tail vane to position the propeller correctly.

A permanent or fixed-mount wind generator should have an effective means of being controlled in high-speed winds. (See the comparative evaluation in Chapter 18.) By the nature of their design, the multi-bladed units have self-limiting propellers. But the large prop units must rely on a manual, mechanical or electro-mechanical brake to slow the propeller, as it cannot be rotated 90° to the wind like the rigging-suspended or pole-mounted units. A good

**Fig. 3-Q.** A good-looking, sturdy installation of a Windbugger on the yacht *Wendy K*.

manual brake is a simple back-up to more sophisticated braking methods.

We recently met a sailor who had fix-mounted a Hamilton Ferris unit over the front of his bow pulpit. He told us that a manual brake, similar to a bicycle hand brake, could stop the propeller when needed. After 6 months of cruising he hadn't, however, installed this brake and was stopping the generator by placing his hand on the hub, between propeller and generator, and applying pressure. There is inherent danger in this method of shutdown, further aggravated by the unit's close proximity to anchor lines and the people handling those lines.

*Pole mounts* are mounted on top of a sturdy, well-supported pole usually installed on the stern. On trimarans we have seen them mounted in the bow of one ama or attached to the bow pulpit, but we still favor the stern for better weight distribution and ease of access. The wind unit can rotate 360° and be left in place and operating while underway. To keep the output wires from twisting as the unit tracks the wind, the current is usually transferred from the generator to the wires by slip rings or brushes, allowing the generator to rotate freely on the pole. Most home-built units, and a few manufactured ones, do not include slip rings. To avoid the annoyance of untwisting the wires, owners often limit the travel of the rotating unit (usually by tying off the vane to the pole). At anchor, this rig is entirely satisfactory.

*Laying the Groundwork*

All large-diameter prop pole mounts must be manually or mechanically controlled in 25–30 knot winds to prevent damage to the generator and/or propeller and frame. They should also include a means of breaking down the entire pole assembly in the event of an emergency. We've rarely seen this. Most pole installations are concocted by owners who find it easier to fix just one piece of pipe to the deck.

Small-diameter prop pole mounts are easier to mount safely (away from random fingers, heads, etc.), do not require attention, and may be left up in any conditions. They are the *only* pole-mounted wind units suitable for use on long offshore passages.

## WATER

The function of a water generator is very similar to that of a wind generator, since both extract energy from a moving fluid. The amount of electrical energy we can extract is dependent on the cube of the boat speed, the size and efficiency of the propeller, the efficiency of the generator and the density of the working fluid (in this case water, which is approximately 780 times denser than air!).

### Classification of Water Generators

BY PROPELLER TYPE AND LOCATION:

*Trailing-log type*—With these water units, the generator usually is suspended in a small gimballed bracket on the port side of the stern. A 3′ or longer stainless steel shaft with a small (about 10″ or less diameter) propeller on the end is trailed behind the boat on 50–100′ of tightly wound braided line. As the propeller and shaft are pulled through the water, they spin rapidly, usually "walking" slightly to port. The braided line transfers the rotary motion to the generator shaft.

A variation on this theme is the Power-Log water unit, which has the generator attached directly to the propeller, like a wind unit. The whole assembly is trailed behind the boat with the propeller motion transferred directly to the generator shaft in the water.

It's worth noting that many wind unit manufacturers also offer a

**Fig. 3-R.** A Hamilton Ferris water trolling unit ready for generating. Note gimbaled frame mounted to the stern that holds the generator (courtesy Hamilton Ferris).

**Fig. 3-S.** Power Log water generator with integral propeller and charging unit connected to tow line trailed behind a boat.

water conversion kit that uses the same generator for both systems. This is an excellent choice for sailors wanting the flexibility of two methods of generating electricity.

*Auxiliary propeller through the hull*—When building a boat, or as an added feature to existing boats, a separate shaft and propeller can be mounted through the hull to turn a generator. If the correct pulleys are used, the shaft will turn at a high enough speed to rotate a high-output alternator. A little design work will be necessary to get the correct prop/alternator/pulley combination. For the power available, it is best to first purchase a special generator made to run at low speed (available from a wind or water-generator manufacturer), then connect it as a direct drive (or 1-to-1 pulley ratio). A small 8″–10″ outboard motor propeller similar to the trailing-log type propellers can be used, resulting in minimal drag and wear on the through hull bearings.

*Auxiliary generator on existing freewheeling propeller shaft*—While the boat is under sail, many sailors are taking advantage of the otherwise unused energy available from a freewheeling propeller. Some companies are offering an alternator and mounting system that is specially designed to operate at the relatively low speed of a free-

**Fig. 3-T.** An example of a generator coupled to an auxiliary shaft through the boat's hull.

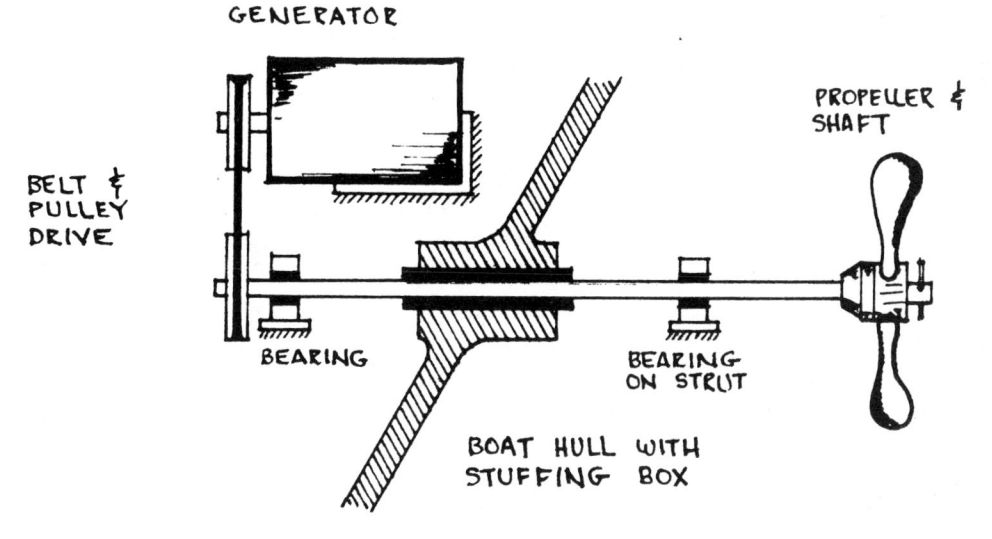

*Marine Alternate Energy Equipment—A Generic Description*

wheeling shaft. Usually included is a voltage regulator to prevent overcharging. A generator, purchased from a wind/water generator manufacturer, also can be fitted by means of pulleys to your shaft. Some conversion kits are available for a propeller shaft that employs the same generator used on a wind or water unit. These types of generators should be electrically isolated from your battery when the engine is running if the shaft speed is excessive. Some means of manual or electrical battery overcharge protection is recommended, either in the form of a voltage regulator or a manual/electrical disconnect switch. There is some controversy over the possible harmful effects of freewheeling on your transmission. Despite statements from manufacturers that most transmissions are not harmed by freewheeling, a method of locking the shaft is a good investment. Locking mechanisms such as Shaft Lock and Proplock are compatible with propeller shaft generators, and can be employed when the batteries are charged or the boat speed is low.

Another controversy in using freewheeling propeller shafts to generate electricity is how this method affects boat speed. Manufacturers of locking mechanisms claim that boat speed is increased by locking the shaft, while some generator manufacturers claim that freewheeling actually increases boat speed.

Our research shows that for most boats, locking the shaft will increase boat speed when sailing at half your boat's hull speed or less. When sailing faster than half of hull speed, freewheeling the shaft more than likely will increase boat speed.

The amperage output while sailing will depend on the size and configuration of the propeller, the type of generator, and the boat speed. This system *will not work* on a boat with a feathering propeller that folds together when the engine is not rotating in gear.

*Outboard leg type*—This type of water generator, which looks very similar to a small outboard motor, is suspended at the stern of the boat. The propeller is immersed in the water where it spins, transferring rotary motion to a generator at the top of the unit.

# PART II
# Assessing Your Needs

# CHAPTER 4

# ELECTRICAL APPLIANCES—HOW THEY DETERMINE YOUR LOAD

The electrical output you need from an alternate energy system is determined by the nature and size of the electrical appliances you have, or plan to have, on board your boat. This chapter will list all electrical equipment usually found on recreational sailing craft, along with the gear's electrical draw in amps and the average number of hours it is used per day. We'll also discuss ways to reduce your electrical demand, thus increasing the feasibility of using alternate energy to handle your load. Finally, let's examine four hypothetical boats, with very light, light, medium, and heavy electrical demand, including a calculation of each boat's average daily load. From this information you will determine a category for your boat to see how future equipment purchases will add to your total load.

The electrical load for each appliance shown below is an average of the products available for boats, and will change slightly with battery voltage. Check your individual equipment to get exact amperage draw. (Remember: Watts = Volts X Amps.)

## WAYS TO REDUCE YOUR ELECTRICAL DEMAND ON A BOAT

Only you can determine which boat electrical appliances you need or would like to have, and which ones you can do without. Before purchasing an AE system, take a close look at your total load to see if there are easy ways to reduce this load without sacrificing your personal feelings of comfort and safety. We are well aware that a few simple luxuries on board a boat can make life much more pleasant. We like to be able to turn on a light or play music any time we feel like it. With children it's nice to not have to constantly worry about whether they turned their light off or how much electricity they are using playing their favorite Peter Pan tape. We sized our AE system accordingly and consequently always have enough electricity. But we do have energy efficient appliances that decrease our load and stretch our electrical supply.

Before installing a $3,500 solar system to supply hot water to a house, one should make sure that $10 water-saving showerheads and faucet nozzles are being used. These significantly reduce the hot water load. Before a passive solar system is designed for a new residence, the house itself must be energy-efficient. Similarly, a boat's energy demand, particularly its lighting and refrigeration load, must be reduced where possible before sizing an AE system, as your initial cost of

## ELECTRICAL LOAD OF MARINE APPLIANCES

| Appliance | Electrical Load (Amps) | Typical Daily Hours of Use | Total Daily Consumption |
|---|---|---|---|
| **Cabin Lights** (Incandescent or fluorescent) | | | |
| 6 Watt | 0.5A | | |
| 8 Watt | 0.6A | 3.5 Hours | Depends on |
| 12 Watt | 1.0A | of cabin | number and |
| 25 Watt | 1.2A | lighting | combination of |
| 25 Watt | 2.0A | each night | lights used |
| 40 Watt | 3.2A | | |
| **Running Lights** | | | |
| Steaming light | 1.0A | 11.0 | 11.0A |
| Masthead tricolor | 2.0A | 11.0 | 22.0A |
| Port/starboard/stern | (1.0A each) | 11.0 | 33.0A |
| **Anchor Light** | | | |
| Small draw (see later this chapter) | 0.05A | 11.0 | 0.55A |
| Regular masthead | 1.0A | 11.0 | 11.0A |
| Bright masthead | 2.0A | 11.0 | 22.0A |
| Spreader lights (2) | 5.0A | | |
| Sealed spotlight beam 300,000 candle power | 12.0A | Intermittent | Depends on use |
| Sealed spotlight beam 200,000 candle power | 7.5A | | |
| Halogen bulb 110,000 candle power | 4.0A | | |
| Strobe light | 0.75A | | |
| **Instruments** | | | |
| Depth sounder | 0.5A | | |
| Knotmeter | 0.1A | Depends on |  |
| Knotmeter w/log | 0.1A | the sailing |  |
| Wind speed indicator | 0.1A | you do |  |
| Wind direction | 0.1A | | |
| Instrument lights | 0.1A | | |
| **Bilge pumps** | | | |
| 400 GPH | 2.0A | 0.5 Hs | 1.0 |
| 800 GPH | 4.0A | 0.5 Hs | 2.0 |
| 1750 GPH | 8.0A | 0.5 Hs | 4.0 |
| 3500 GPH | 15.0A | 0.5 Hs | 7.5 |
| Water pressure pump | 4.0A | 0.5 Hs | 2.0 |

| Appliance | Electrical Load (Amps) | Typical Daily Hours of Use | Total Daily Consumption |
|---|---|---|---|
| Shower pump (on demand) | 2.0A | 0.5 Hs | 1.0 |
| Seawater intake pump (clear) | 4.0A | 0.5 Hs | 2.0 |
| Seawater intake pump w/barnacles covering the opening | up to 15.0A! | 0.5 Hs | 7.5 |
| **Electronics** | | | |
| Stereo | 2.0A | 3.0 Hs | 6.0 |
| Tape deck  large draw | 2.0A | 2.0 Hs | 4.0 |
|   small draw | 1.0A | 2.0 Hs | 2.0 |
| AM-FM receiver | 0.5A | 3.0 Hs | 1.5 |
| TV  small B & W | 1.25A | 2.0 Hs | 2.5 |
|   small color | 4.0A | 2.0 Hs | 8.0 |
| CB  receiver | 0.5A | 2.0 Hs | 1.0 |
| VHF  receiver | 0.5A | 2.0 Hs | 1.0 |
|   transmitter low | 1.0A | 2.0 Hs | 2.0 |
|   transmitter high | 5.0A | 0.5 Hs | 2.5 |
| SSB  receiver | 1.0A | 0.5 Hs | 0.5 |
|   transmitter | 20.0A | 0.5 Hs | 10.0 |
| Loran  standard | 0.5A | 4.0 Hs | 2.0 |
|   with memory | 1.0A | 4.0 Hs | 4.0 |
| Sat-Nav  standard | 0.5A | 4.0 Hs | 2.0 |
|   with memory | 1.0A | 4.0 Hs | 4.0 |
| **12V Motors** | | | |
| Refrigerator compressor | 5.5A | 10.0 Hs | 55.0 |
| Autopilot | 0.5A | 12.0 Hs | 6.0 |
| Anchor windlass | 20.0A | 0.1 Hs | 2.0 |
| Engine compartment blower | 2.5A | 1.0 Hs | 2.5 |
| Cabin fan (oscillating) | 1.2A | 4.0 Hs | 4.8 |
| 12V blender | 2.5A | 0.5 Hs | 1.25 |
| **12V Tools** | | | |
| Drill | 10.0A | Depends on the work you do | Depends on the work you do |
| Saber saw | 15.0A | | |
| Soldering iron small | 5.0A | | |
| **Misc.** | | | |
| Horn | 2.0A | 0.5 Hs | 1.0 |
| Bell | 1.0A | 0.5 Hs | 0.5 |

equipment will be correspondingly less. Let's take a look at the major areas of concern and see how we can make a boat more energy efficient.

## LIGHTING

### Cabin Lighting

Even though the warm glow from a kerosene lantern is hard to beat for atmosphere, electric lighting is better for reading (a major source of nighttime entertainment), cooking in the galley (what *was* that I just threw in the pot?), and safety especially with children and pets). In addition, if you cruise tropical areas or in the heat of summer, that warm glow is going to be less appreciated. Our Tilley kerosene lantern with a mantle provides good light and heat on chilly nights but becomes stifling in warm weather.

By using fluorescent cabin lighting you can reduce your electrical lighting load by 80%. Lighting efficiency is measured in lumens per watt, or the light output in relation to the energy consumed. A typical 8W fluorescent cabin light, delivering 60 lumens/watt will give the same light output, or lumen level, as a 40W incandescent bulb yielding 12 lumens/watt. It also generates less heat.

Incandescent lights operate similarly to resistance-type heating elements. A tungsten filament is suspended between 2 metal contacts inside a vacuum bulb. As electricity is supplied, the resistance of the filament determines the current. Only 10% of the electricity consumed is turned into light (radiation energy) while 90% is dissipated as heat. This is the reason an incandescent bulb gets so hot.

Fluorescent lights operate quite differently. They use a starter at the end of the fixture to spark an electrical charge that flows through cathodes in either end of the fluorescent tube. This charge excites gaseous atoms inside the tube. As they discharge, they cause the phosphor coating on the inside of the tube to fluoresce emitting a large quantity of visible light.

We've heard it said that fluorescents give off a "different" light than incandescents, one not suitable for cozy living areas. Incandescent lights bring out the warm tones, red, orange and yellow, while fluorescent bulbs produce cool tones, blue and green. This is called the color rendition. A bare fluroescent bulb can be a little harsh on the eyes but most fixtures you purchase will have a plastic cover that diffuses light. If you still find the lighting uncomfortable, you can recess it by placing a wooden frame set away from the fixture so the light is reflected up and down against the wall but is not directly visible. This technique is very effective and attractive.

**Fig. 4-A.** Reflective lighting technique for fluorescents.

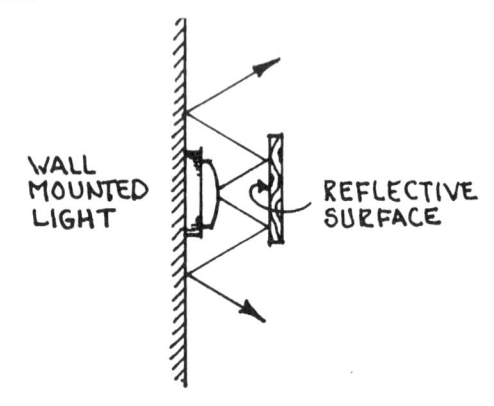

*Assessing Your Needs*

On *Kjersti* we have an 8W fluorescent fixture in each aft cabin and one in the main cabin for general lighting. There is also a 6W fixture in the head and over the galley. Since the light in the main cabin is at the opposite end of the settees from where we like to read, we supplement it with a portable 7W fluorescent lantern (attached to a cigarette lighter plug) that hangs over our reading area. It cost us about $10 at a discount store and comes with a flashlight and emergency flasher. The flashlight is convenient as we can do without throwaway batteries. There is also a 10W high intensity incandescent bulb over the navigation/drawing board. If you have any incandescent bulbs on board, for reading or work lights, they should be the high intensity type with reflective shades that direct light where you need it, and thus use less energy. Ours uses a bulb found in foreign car taillights. These fixtures give us ample levels of light. On a typical evening with one of us at the drawing board, the children in their cabin, and the other in the galley, our total lighting load is:

8W (children) + 6W (galley) + 10W (drawing board) +
8W (main cabin) = 32W or a 2.6 amp draw.

This is about the same as one 30–35W incandescent bulb.

A few other things you do can further increase the output of your lighting fixtures.

- In the main cabin and head, we cut a narrow 1″ slit in the bottom of the plastic cover on the lighting fixtures. This

**Fig. 4–B.** Slot in light cover increases efficiency of bulb without increasing glare.

allows more light to escape without having to look directly at the bulb.
- In the navigation/drawing board area we removed the plastic cover on the high intensity light and installed a reflective shield behind the bulb, directing light right onto the drawing area. Most high intensity light fixtures (also called reflector or directional lamps) come with a reflective shade to achieve the same effect.
- Keep bulbs and fixtures as clean and dust-free as possible. A layer of dust on a bulb can easily cut the light output in half.
- An incandescent bulb will lose 20% or more of its rated light output near the end of its life. If you use these bulbs, replace them when they first become dim.

## ANCHOR LIGHTS

An anchor light not only protects you from collision, but it aids other boats entering or leaving an anchorage at night by making you visible. Having sat out a gale in a tight anchorage surrounded by unlit boats, making it impossible to see if they were dragging (and several did), we have developed strong opinions on the subject of anchor lights. On boats equipped with an electric anchor light, not turning it on seems to be a favorite energy-saving device. This practice is dangerous, irritating and unnecessary.

Most electrical masthead anchor lights are a big battery drain. They consume 10 amp-hours (10W bulb) to 25 amp-hours (25W bulb) in one night. While kerosene lights are a good alternative, they still use a non-renewable oil product and many boats with inexpensive lamps can't use them in any wind as the flame blows out.

An energy efficient electric anchor light can easily be made that supplies the necessary light and utilizes your AE system with a fraction of the electrical load of other lights.

**Fig. 4–C.** A low drain anchor light you can make onboard.

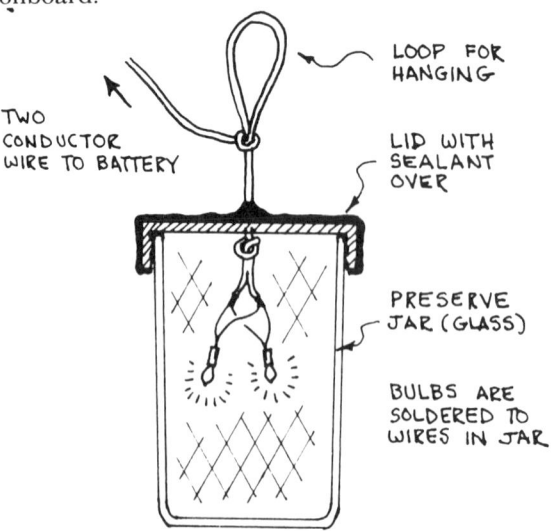

We made one that gives off as much or more light than a kerosene lantern, uses only 50 milliamps (ma) of electricity (equivalent to 0.6 amp-hours for 12 hours of use), and costs about $4. See Fig. 1–C.

### Low Drain Anchor Light

Take a small, fractured glass preserve jar with a metal canning lid. Punch a hole in the center of the lid big enough to pass through a 2-conductor #14 gauge insulated wire (exterior use). Insert this wire through the lid and tie a knot on the inside. Bare the ends of the two leads. Solder 2 high-intensity panel bulbs. Each rated at 25 ma, 12V DC (available at Radio Shack with wires attached for 79¢ each) in *parallel* to the #14 gauge wire. Rustproof the metal lid with silicon, making sure to seal the area where the wire passes through. Make a loop in the wire outside the jar to attach a small lanyard to hoist the light to the spreaders. If a brighter light is desired, bulbs with higher amperage ratings, and therefore higher light intensity, may be substituted. This anchor light is easy to use and needs only an occasional airing out to release any trapped condensation. We use ours on a regular basis, reserving the brighter masthead anchor light for the few times we anchor in exposed areas, or near active night boat traffic.

### Photoelectric Switches

Kuau Technology, Ltd. of Maui, Hawaii, makes a line of marine-grade photoswitches that automatically turn lights on at sunset and off at sunrise. When applied to anchor lights this device is very economical and con-

**Fig. 4-D.** A photoswitch used to automatically control anchor light and other accessories (courtesy of Kuau Technology Ltd.).

venient. It helps avoid any unnecessary battery drain if you forget to turn your light off. If you have a small solar panel it can easily power a low-drain anchor light similar to the one described earlier. The combination of a solar-powered light and switch will put an end to anchor light headaches and battery concerns. On unattended boats the switch takes care of the anchor light requirement, a very real problem in places like Vero Beach, Florida, where there is a $50 penalty for failing to use an anchor light. That's about $10 more than the cost of the automatic switch. Kuau Technology's address is: P.O. #1031 Puunene, Maui, HI 96784.

## RUNNING LIGHTS

Amazing as it may seem, we have met sailors who conserve on electricity at the expense of running lights. The risk involved is irrational, since even a modest AE system can easily take care of this load. Installing a masthead tricolor using only a single 25W bulb instead of 3 separate ones offers a significant energy savings. Leave your deck-mounted running lights in place on a separate switch so you can use the lower set when it seems safer. If you would rather not use electricity, then a *good* quality set of kerosene running lights can be substituted. But if your lights are electrically operated, it is a sailor's obligation to use them when traveling at night.

## REFRIGERATION

Although most of the world lives without it, refrigeration is a wonderful convenience that many Americans desire on their boats. Doing without is no great hardship, but when refrigeration is available we all appreciate it, even if just to cool a beer on a hot day. In the Bahamas, ice cubes and cold drinks are a prime trading commodity. Anyone who could supply one or both was a welcome guest at any party, if not the cause of the party in the first place. When ice is available, an ice box can provide a reliable, fairly inexpensive means of refrigeration. But many people don't enjoy the hassle of finding ice or the mess of dealing with it.

As soon as you install a 12V refrigerator on your boat, your electrical load will soar. Only a few AE systems offer outputs capable of sat-

*Electrical Appliances—How They Determine Load*

isfying this load, and then only in ideal conditions. Despite refrigeration's high energy consumption, its susceptibility to breakdown, and difficulty of repair, it is becoming increasingly popular.

## The Basics

There are two types of 12V refrigeration, thermoelectric and compression cycle.

*Thermoelectric* refrigerators use solid-state circuitry. With no compressor to fail or freon gas to escape, they are very reliable. The ones you see advertised are good for cooling a small volume of food in temperate climates. In the tropics they have to run most of the time to keep things cool. They also have no freezer compartment. This relatively new device is now available in small portable units. One of the better known thermoelectric refrigerators, the Koolatron, draws 4 amps when running. When it's 85° outside, the cooler must run 12 hours out of 24 to keep 30 lbs. of food cold. Koolatron also offers a thermoelectric cooling module for use in your own icebox where much greater insulation values are possible, thus less running time. See Fig. 4–E.

*12V Compression Cycle Refrigerators* are similar to home refrigerators, using a 12V motor-driven compressor in a vapor-compression cycle to pressurize the working fluid (usually Freon 12). The gas travels between the evaporator, where heat is drawn from the space to be cooled, and condensor, where this excess heat is given off to the environment. The compressor is the main source of electrical draw, the other being the light on the inside when you open the door or lid. Why not remove the light since you really don't need it.

There are many different makes of boat refrigerators. Some come as a complete unit, like a small refrigerator, while others have a separate cooling unit that can be installed in any ice box compartment. The compressor/heat exchanger of these units often are installed in the engine compartment.

## The Electrical Load

One marine refrigerator manufacturer quotes that a well insulated moderately sized refrigerator, to provide the equivalent of 10 lbs. of ice a day, will use 37 amp-hours at 12V. Most people we've talked with, however, say their

**Fig. 4–E.** Koolatron cooling module; a thermoelectric device for use in an insulated icebox (courtesy of Koolatron).

fridges draw approximately 5 amps when running and are on about 8–10 hours a day (40–50 amp hours) in moderate weather and up to 12–14 hours a day (60–70 amp-hours) in hot weather. This means that an alternator putting out 15 amps would have to be run 3 hours for the former and 4-½ hours for the latter. You can check your manufacturer's specs to calculate the load of your own refrigerator.

## Ways of Reducing the Load

How can you increase the efficiency of your refrigerator?

The current draw of a 12V refrigerator, or any electrical ice box, depends on the following things:

*Outside (Ambient) Temperature*—There's not much you can do about that except cruise in cool places. You can reduce the cabin temperature where the fridge is located by scrupulous use of a large awning and adequate wind scoops.

*Refrigerator Location*—Locate it so the sunlight does not fall directly on the refrigerator's surface. If this isn't possible place a light colored towel or equivalent over the top to reflect the sun. A damp towel will also help cool the box due to evaporation. Place the refrigerator as far as possible from any heat source.

*Inner Compartment Size*—Make the box as small as you can. There's no point in using energy to cool space that is never used. If you are a once-a-week shopper, see how much space you require to refrigerate all your perishables at the beginning of the week, then limit your refrigerator to that size. If your fridge is larger than you need, line the inside with an inch or two of styrofoam board (not urethane, which eventually will absorb moisture) to reduce the size and increase the box's efficiency. Many foods that find their way into a refrigerator don't need to be there. Americans have a habit of using a refrigerator like a pantry, loading it with fresh foods and leftovers that really don't need to be cooler than room temperature. Eggs, most produce, mayonnaise, butter and most leftovers, to name a few, keep very well simply stored in a cool part of the boat. And just because you *have* a case of beer doesn't mean it all has to be chilling at once.

*Thermal Blanket*—To further increase the efficiency of your refrigerator, a lightweight thermal blanket or "space blanket" can be laid over the food to hold in the cold. Make sure you also lay the blanket over the cooling unit or it won't do any good. This effectively adjusts the space to be cooled to the amount of food you have.

*Inner Compartment Temperature*—Your electrical load will be affected by how cold you like to keep your refrigerator. The colder the fridge, the more electricity it uses. If yours makes ice, how much electricity it draws will correspond to how much ice you need. Ice water may be free in American restaurants, but in terms of electricity, it's very expensive on your boat.

*Exterior Insulation*—In real estate they say the three most important selling points of a house are location, location and location. In like fashion, the three least expensive, easiest and most effective ways of reducing your refrigerator's electrical load are to insulate, insulate, insulate. This includes not only the inside compartment but also the cover. The fridge should have a completely airtight seal around the opening. Most manufacturers do *not* insulate properly. One sure way to tell if yours is poorly insulated is to look for condensation or "sweat marks" on the outside of the hull or interior where the refrigerator is located. This is cold escaping and hitting a warm surface. It would not be overkill to have up to 6" of rigid urethane foam insulation completely surrounding the box, and a well-insulated lid.

*Access Door Location*—The most efficient door is located on top of the refrigerator, on boats usually designed as a counter cutaway. Side-opening doors allow cold air sitting in the lower part of your refrigerator to "fall out" when the door opens. This is the reason refrigerators in homes almost always kick on when the door is opened.

*Frequency of Use*—Schedule your door openings so you can get out or return everything you need at one time. At mealtimes plan what you want from the fridge, open it once and remove everything quickly. A refrigerator that is left open or opened frequently must use that much more electricity to keep it cool. Always allow heated food to cool completely before placing it in the refrigerator. Don't leave cold things out until they warm, for the more food at room temperature you place in the fridge, the more energy must be used to cool it.

*Compressor Unit*—Try to install the compressor in a cool, well-ventilated space to increase its efficiency.

## AE Systems For Boats With 12V Refrigeration

You may already have heard conflicting reports of what type of alternate energy system is required for boats with 12V refrigeration. This conflict arises from misconceptions regarding the electrical output of various AE systems and from the wide variation in electrical loads imposed by 12V refrigeration.

It's possible to power a refrigerator with only two 35W solar panels, just as it's possible to own a high output wind generator and still not satisfy your refrigeration load. In order for two solar panels to do the job, your cooling load would have to be half of the 50–60 amp-hours typical for 12V boat refrigeration. In addition, the rest of the electrical load must be considered when selecting an AE system to handle refrigeration. But the fact that solar panels can, and have, successfully powered refrigerators serves as a good example of what can be done when people practice conservation both in their choice of refrigeration and in the handling of their cooling load.

A more typical daily electrical load for a boat with moderately efficient refrigeration might be 50 amp-hours for the refrigerator and 25 amp-hours for additional electrical needs, or a *total* of 75 amp-hours per day. As a rule, two kinds of AE systems can potentially satisfy this total daily load on a regular basis; large-diameter prop wind generators and high output water generators. Refer to Chapter 9 for a rough estimate of the average daily output for these individual systems. It must be remembered that if generating conditions are not adequate, the output of these systems will drop and a supplementary power source will be required.

A popular alternative to 12V refrigeration is an engine-driven system where the compressor is run directly off the engine shaft, similar to a car air conditioner. We have mixed feelings about this type of system. On the one hand it will *only* operate by a daily running of the engine, thus denying any chance for alternate energy to help with this mechanical load. This also imposes an additional load on the engine. On the other hand, the engine running time with this system is usually much less than the charging time required for a 12V refrigeration system. And, the alternator can be charging your batteries at the same time. An AE system could then easily handle the balance of your electrical load. If we were to install refrigeration on a boat with an engine-driven alternator, we would have a small, super-insulated ice box with a 12V motor-driven, compressor-type cooling unit. We would employ a hybrid alternate energy system (see Chapter 8). Finally, we would install a manual control to boost the output of the engine's alternator for the few times supplementary charging was necessary.

If you still are undecided about refrigeration, try installing a well-insulated ice box first. We lined our small portable cooler with 1″ styrofoam, including the inner lid, added a reflective foil face to the top, and covered it with two blankets. In 80–90° weather we found a 10 lb. block of ice lasted 3 days. The same can be done to a standard built-in ice box. By following the tips listed in reducing your refrigeration load, you might find ice to be quite satisfactory. Without refrigeration, your AE system will easily cope with your electrical needs, provided you have sized it correctly. With refrigeration, you will have to carefully size your system, conserve where possible, and seek optimum generating conditions.

## ELIMINATING DISPOSABLE BATTERIES

Once you own an AE system that produces free, renewable electricity, you can convert most of your battery-operated appliances to 12V. Not only is this better environmentally (nothing to throw away) but it frees you from the cost and worry of dead batteries.

Battery-operated appliances that can be converted to 12V include tape decks, radios, flashlights, lanterns, spotlights, anchor lights, RDFs, calculators, etc. If your appliance alrady has a 12V adapter, it can easily be plugged into a 12V outlet (also known as a cigarette lighter outlet) that is powered by your boat battery. Most battery-operated appliances operate on less than 12V, determined by the voltage and number of batteries (cells) connected in series.

### Examples

| | | | |
|---|---|---|---|
| (2) | "C" size at 1.5V each | = | 3V |
| (4) | "AA" size at 1.5V each | = | 6V |
| (4) | "D" size at 1.5V each | = | 6V |
| (1) | 6V "lantern" battery | = | 6V |

*Electrical Appliances—How They Determine Load*

**Fig. 4-F.** A typical power transformer 12V DC into 3, 4, 5, 6, and 9V DC.

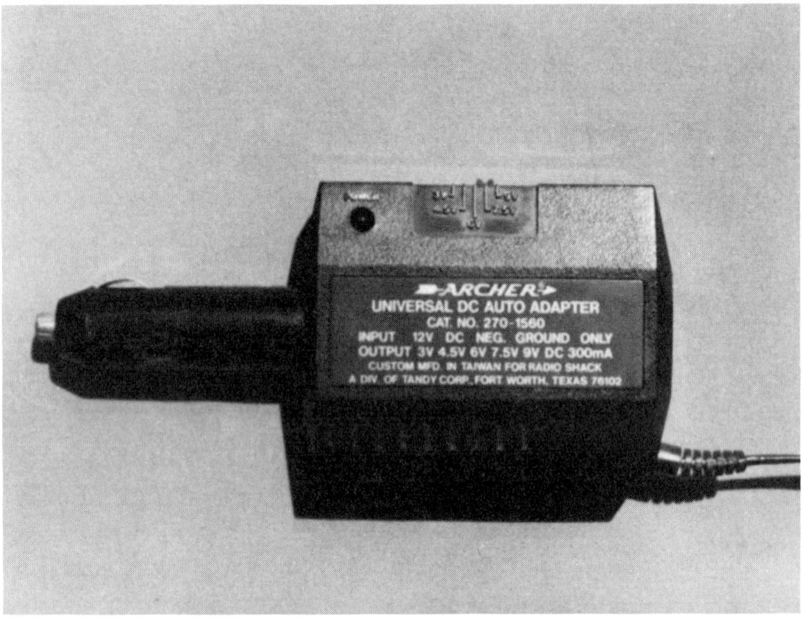

| | | |
|---|---|---|
| (6) | "D" size at 1.5V each | = 9V |
| (1) | 9V radio battery | = 9V |

If your appliance does not come with a separate 12V DC adapter, purchase an Archer Universal DC Auto Adapter from Radio Shack. This plugs into your 12V outlet and has a selection of voltages to choose from, depending on your appliance. You will also need an adapter cord to the appliance. We soldered this cord to an internal connection on our SW/AM-FM broadcast radio that allows operation on either 4 "D" cells or 12V (transformed to 6V) from our boat battery.

## RECHARGEABLE BATTERIES

While the initial cost is high, the Ni-Cad (nickel-cadmium) rechargeable cells are an attractive alternative for small loads. There is a unit available that will recharge them from your 12V boat battery, thus making use of your AE system. This might be especially practical for remote load, such as battery-operated anchor lights or radio/tape decks that are taken ashore. See Solarex, Energy Sciences Catalog, in Chapter 17.

## EXAMPLES OF AVERAGE DAILY ELECTRICAL LOAD

The following is an examination of our four hypothetical boats with electrical loads ranging from very light to heavy. Included is the electrical demand of the individual appliances plus a calculation of each boat's average daily load.

*Assessing Your Needs*

## BOAT WITH A VERY LIGHT LOAD

| Appliance | Total Power Required | Current (Amps) Draw | Daily HRS of Operation | Total Daily Load (Amp-Hours) |
|---|---|---|---|---|
| Cabin lights | 16.0W | 1.33 | 3.0 | 4.0 |
| Small draw anchor light | 0.6W | 0.05 | 11.0 | 0.6 |
| AM/FM/SW receiver | 2.0W | 0.17 | 2.0 | 0.34 |
| VHF    receive | 6.0W | 0.5 | 0.17 (10 min.) | 0.09 |
|          transmit | 60.0W | 5.0 | 0.02 (1 min.) | 0.1 |

= 6 Amp-Hour/Day

## BOAT WITH A LIGHT LOAD

| Appliance | Total Power Required | Current (Amps) Draw | Daily HRS of Operation | Total Daily Load (Amp-Hours) |
|---|---|---|---|---|
| Cabin lights | 24.0W | 2.0 | 3.5 | 7.0 |
| Small draw anchor light | 0.6W | 0.05 | 11.0 | 0.6 |
| AM/FM/SW receiver | 2.0W | 0.17 | 3.0 | 0.51 |
| Tape deck/TV | 15.0W | 1.25 | 2.0 | 2.5 |
| VHF    transmit | 60.0W | 5.0 | 0.05 (5 min.) | 0.25 |
|          receive | 6.0W | 0.5 | 0.5 (30 min.) | 0.25 |
| Running lights (one night every two weeks) | 25.0W | 2.0 | 0.86 (daily average) | 1.72 |
| Small bilge pump | 48.0W | 4.0 | 0.33 (20 min.) | 1.32 |

= 14 Amp-Hrs/Day

*Electrical Appliances—How They Determine Load*

## BOAT WITH A MEDIUM LOAD

| Appliance | Total Power Required | Current (Amps) Draw | Daily HRS of Operation | Total Daily Load (Amp-Hours) |
|---|---|---|---|---|
| Cabin lights | 36.0W | 3.0 | 4.0 | 12.0 |
| Anchor light | 12.0W | 1.0 | 11.0 | 11.0 |
| Running light (once every 11 days) | 25.0W | 2.0 | 1.6 | 3.2 |
| Tape deck/TV | 15.0W | 1.25 | 3.0 | 3.75 |
| AM/FM/SW receiver | 3.0W | 0.25 | 3.0 | 0.75 |
| VHF   receive | 6.0W | 0.5 | 2.0 | 1.0 |
|       transmit | 60.0W | 5.0 | 0.5 | 2.5 |
| Bilge pump | 48.0W | 4.0 | 0.5 | 2.0 |
| Loran | 6.0W | 0.5 | 4.0 | 2.0 |

= 40 Amp-Hrs/Day

## BOAT WITH A HEAVY LOAD

| Appliance | Total Power Required | Current (Amps) Draw | Daily HRS of Operation | Total Daily Load (Amp-Hours) |
|---|---|---|---|---|
| Cabin lights | 48.0W | 4.0 | 3.5 | 14.0 |
| Anchor light | 12.0W | 1.0 | 11.0 | 11.0 |
| TV/tape deck | 15.0W | 1.25 | 4.0 | 5.0 |
| SSB   receive | 12.0W | 1.0 | 3.0 | 3.0 |
|       transmit | 240.0W | 20.0 | 0.4 | 8.0 |
| Autopilot | 6.0W | 0.5 | 12.0 | 6.0 |
| SAT-NAV/Loran | 6.0W | 0.5 | 4.0 | 2.0 |
| 12V refrigeration (well insulated) | 60.0W | 5.0 | 10.0 | 50.0 |
| Anchor windlass | 240.0W | 20.0 | 0.05 | 1.0 |

= 100 Amp-Hrs/Day

# CHAPTER 5

## THE RIGHT ALTERNATE ENERGY SYSTEM FOR YOUR BOAT

It's easy to become overwhelmed with today's choice of boats. The market is full of everything from small daysailers to spacious cruisers, from sleek, ultra-light racers to heavy-displacement, traditional wooden craft. Some boats sprout forests of rigging while cat ketches boast no standing rigging at all. Some boats have one hull, others three. Every boat has its own characteristics, appeal and alternate energy potential. The purpose of this chapter is to help you decide which AE system(s) will work best on your particular type of boat. All the basic types of sailing craft plus some pertinent design features will be discussed with respect to how they affect a specific alternate energy system.

Two points are well worth mentioning. First, any boat with a power load, no matter how small, can benefit from a solar, wind or water electrical generating system. Just because your energy consumption is slight is no reason to disregard alternate energy. On the contrary, a boat with only a small electrical drain is ideally suited for a small, inexpensive source of alternate energy to supply *all* its electrical needs.

Second, when selecting a new boat, look at its potential to generate alternate energy. Although it may not be as important as seaworthiness or appropriate accommodations, the boat's adaptability to a renewable energy system should be a part of your decision.

When considering an alternate energy system, boat size exercises the greatest influence over your choice. It determines the amount of space you have to work with and often indicates the amount of energy that needs to be generated.

Beginning with dinghies, we will propose a variety of alternate energy systems that are most applicable to different-sized boats.

## DINGHIES

Motorized dinghies rely on 2-cycle gasoline outboard engines, with all their inherent noise, pollution, cost, maintenance and annoyance. Motorized dinghies have done for quiet anchorages what lawnmowers have done for quiet weekends at home. However, there is a practical and energy-efficient substitute for the gasoline engine. No, we don't mean oars. If you are using an outboard you likely have already tried and rejected rowing. The alternative is an electric trolling motor. It is reliable, relatively inexpensive, lightweight, clean, quiet and capable of being

**Fig. 5–A.** A solar powered dinghy! All you need is an electric trolling motor, a battery, and a photovoltaic solar panel.

powered by renewable energy. We have seen several dinghies with electric motors run by a standard 12V boat battery that was recharged periodically by the marine engine alternator.

The major advantage of a small electric motor is that it immediately presents the option of using alternate energy to charge your battery. Although it seems like much more, most dinghies are only used a maximum of 30 minutes a day (2 roundtrips at 15 minutes each). A typical electric trolling motor will draw 15A × 0.5 hours/day to equal 7.5 amp-hours/day. At that load a battery could be recharged by one or more alternate energy systems. Another option is to follow the lead of a friend of ours who mounted a 35W solar panel on the bow of his dinghy and a 12V battery under the seat. With moderate use of the motor the solar panel keeps the battery charged.

During the summer on Cape Cod, our friend uses his dinghy to cross the harbor to his moored yacht four or five times a week.

If you would like to try a similar arrangement with your dinghy, Sears carries a good line of electric trolling motors. Another high quality unit is the Minn Kota, 1531 Madison Ave, Mankato, MN 56001. These motors are rated by pounds of thrust. Manufacturers claim that a pound of thrust should be used for every 100 lbs. of boat, gear and passengers. Even if you double that figure to be on the safe side, a 15 lb. or 20 lb. thrust trolling motor is sufficient. Since the trolling motor is simply a permanent magnet motor capable of either using electricity or generating it, you might even consider modifying it for some minimal water or wind generating when it's not in use on the dinghy!

In testimony to the feasibility of electric-powered dinghies, we recently took a ride on a 7-ton Southern Cross 31 that uses a large 12V DC motor for an auxiliary engine. The motor is used mainly to get the boat in and out of harbor. Two 40W solar panels are mounted on deck to keep the batteries charged. The boat glided noiselessly around

60
*Assessing Your Needs*

our anchorage with surprising speed. If the principle works on a heavy displacement cruising boat, it surely can be adapted to dinghies and other small craft.

## DAYSAILERS

While most daysailers (boats without overnight facilities) have no engine or charging system, some do carry a few instruments, such as a radio, running lights or an outboard engine with electric start. All these can easily be run by an alternate energy system. A daysailer requires a system that can adapt to limited mounting and stowage space, that is suitable for sporadic use, and can be left unattended frequently.

*Solar*—Our first choice is a small 7-10W panel that will take up little space and supply all power for an average daysailer (including battery losses). If the boat is equipped with an outboard engine, the solar panel costs the same as an alternator option. or it can be used as a back-up system for days when the sailing was great and the engine was not used.

*Wind and Water*—Both systems are probably either too costly and bulky or their output too high for practical use on a daysailer.

## BOATS SMALLER THAN 28′

Cruising sailboats under 28′ usually are sloop rigged, possess limited stowage space and deck space and usually have a light electrical load, compared to larger boats.

*Solar*—As seen already in our discussion of dinghies and daysailers, no sailboat is too small to accommodate solar panels. Depending on your particular electrical load, you can use one 10W panel, a 20W panel or two 10W panels for easier mounting. If deck space is limited, then you can try mounting them on the stern pulpit. Although one, two or even three 10W panels seem the most practical choice on a small cruiser, the larger panels are comparatively less expensive per watt. Even a 12″ wide by 4′ long 30-35W panel will fit on board if you leave it free to be moved around the boat for maximum exposure when at anchor or on calm days. Add a simple polarized plug and it can be unplugged and stowed below on a bunk when the wind pipes up. We do this on our 26′ catamaran. The trick is to remember to unplug the wire.

*Wind*—While most sailboats under 28′ are too small for pole-mounted, large-diameter prop units, the LVM 25 and 50, and Ampair 100 (36″ diameter prop) will fit nicely, mounted on a pole at the stern or up the mast. On sloop-rigged boats the foretriangle provides a place for rigging-suspended wind units. The one possible problem with the hanging type is stowage when underway. If operating and stowage is a problem, you can consider suspending a small-diameter prop wind unit in the foretriangle.

*Water*—Water generators are an excellent means of generating electricity on a small boat. They take up very little deck or stowage space, are light, easy to handle and have a high output. The Aquair 50 and 100 water generator have a wind attachment, thus supplying two alternate energy options to a small boat with a light electrical load. We have a similar arrangement with our Hamilton Ferris combination wind and water generator.

## BOATS LARGER THAN 28'

For boats over 28' the choice of an alternate energy system is a wide one. Deck or pulpit space for permanently-mounted solar panels is usually available, as is stowage space for a wind or water system. The stern will be large enough to hold a pole-mounted wind generator and the slight drag of a water generator or the additional windage of a mast-mounted wind unit will have little effect on your boat's performance. The choice of solar, wind or water systems will depend more on individual preference and the following considerations: rig, hull configuration and the location of equipment, such a dinghies, Biminis, etc.

### Sloops

Sloops are ideal boats for nearly all types of alternate energy devices.

*Solar*—A solar panel, or panels can be mounted to the large forward cabin area or on the stern rail, where no overhanging boom blocks the sun.

*Wind*—Sloops generally have enough foretriangle space for mounting large, pivoting, rigging-suspended wind generators. Because the backstay on a sloop prohibits excessively long booms, a pole-mount generator can sit on the stern behind the boom, sufficiently out of the way of any sailing activities. A small unit, like the Ampair or LVM, can be permanently mounted on a boat with a fractional rig at the masthead or on the upper forward part of the mast. The upper forward part of the mast on a masthead rig can be used if the head of the jib sets well below the masthead. A large-diameter prop unit can be attached to the lower forward part of the mast above head height if: it doesn't interfere with halyards, spinnaker tracks, etc., if it is accessible for stopping and starting, and if it is not operated under sail when sheets and sails could become entangled in the unit.

### Cutters

*Solar*—A club-footed staysail can block a solar panel permanently mounted forward of the mast. To avoid this, the panel can always be removed from its mount while at anchor and directed toward the sun, thus greatly increasing its efficiency.

*Wind*—With the smaller inner foretriangle of a cutter, you have to determine whether there is enough swinging room for a large rigging-suspended wind unit. Usually there is. Although they are designed to swing 360° on swivels, a smaller area can be achieved by tying off the wind generator so it can only pivot 45° to each side. There still must be a provision to pivot the unit 90° to the wind for shutdown, unless a *good* braking system is used. Our unit on *Kjersti* worked fine until we installed an inner forestay. To solve the problem we now unhook the stay where it attaches to the deck (easy and safe at anchor) when the wind unit is in use. A pole-mounted

*Assessing Your Needs*

wind generator at the stern is fine, providing the boom does not overhang the transom. All smaller wind units fit easily in the available space on a cutter.

## Ketch/Yawl/Schooner

*Solar*—The one thing to keep in mind when mounting solar panels on two-masted boats is the additional shading caused by the extra spars, or sails.

*Wind*—The mizzenmast on a ketch or yawl or mainmast on a schooner are ideal places for permanently-mounted wind units. They are comparatively free of halyards, whisker pole tracks, rings or cleats, but also have no jib to entangle the propeller (take care with mizzen staysails). A wind generator can be either fixed in position facing forward, similar to many Wincharger mounts, or even better, mounted on a short pole similar to a radar mount. Place unit as high as possible to avoid any interference from the mainsail (or foresail on a schooner). Pole mounts are not often seen on these boats because of overhanging booms. We've met several sailors, however, who simply lower their pole and generator out of the way when under sail. Rigging-suspended units would create no problems for a ketch, yawl or schooner.

## Catboats

*Solar*—Both traditional and newer catboat designs have relatively little rigging and clutter, making it easy to find unshaded space for one or more solar panels.

*Wind*—Traditional catboats, with their virtually unstayed, stocky, wooden masts and no jibs, can use a small wind unit mounted to the forward part of the masthead. Without a foretriangle there is no possibility for a rigging-suspended wind unit there, but one can be accommodated in the main triangle. The absence of jibs also makes a pole-mounted unit on the bow pulpit more feasible (see Fig. 5-B).

The newer breed of catboats with completely unstayed masts and wishbone, or half-wishbone rigs, presents another set of considerations. The masts are constructed of aluminum or carbon fiber and should not be

**Fig. 5-B.** Pole mounted wind generator on the bow of a catboat (courtesy of Windbugger).

*Selecting the Right Alternate Energy System*

tampered with. Much of their strength can be attributed to a lack of winches, cleats, fittings, spreaders and fittings that require the drilling of holes, which weaken the mast. The masts also are not designed to function with additional weight and windage aloft. When we asked Rick Strand of Tillotson-Pearson, Inc. his opinion on the subject of mast-mounted wind units, he stressed that large units should *not* be fixed to unstayed rigs. Small wind units could be accommodated, but the designers would have to be consulted first to ensure structural integrity. The best advice is to leave the mast alone and find another mounting location.

## GENERAL HULL CHARACTERISTICS

A few specific hull designs or features can affect your choice of alternate energy system.

*Multihulls*—At the risk of sounding subjective, multihulls are really a superior craft for alternate energy. Our 26′ Heavenly Twins catamaran has deck space for two permanently-mounted 35W solar panels that never are walked on. Her foretriangle accommodates a large-prop wind generator and her 11′ wide stern supplies enough room for a water generator, taffrail log, self-steering and two transom-hung rudders. There is also storage space for the wind and water systems when not in use.

With that amount of room on a small catamaran, you can imagine the space available on larger multihulls. Trimarans have an even larger deck area to work with. We've seen pole-mounted wind units in the bow of one ama. One modern Newick tri had two 35W solar panels mounted side-by-side on the sloping transom.

Water generators are also an excellent choice, as there is ample room to mount the generator bracket and their output is excellent due to the generally superior speed of a multihull. We'll discuss the importance of hull speed later in this chapter.

*Motorsailers*—Like multihulls, motorsailers have wide sterns and plenty of deck and stowage space, making them ideal for a wide choice of systems. In addition, the high cabin roof provides an excellent site for fully-exposed yet unobtrusive solar panels.

### Design Characteristics

*Beam*—Beamier boats will provide more suitable space for solar panels or water generators.

*Transom*—With a double-ender or canoe transom, you will need a small outrigger if you intend to use a water generator in conjunction with a taffrail log or fishing line. An outboard-leg water unit is probably out of the question, particularly if you already have self-steering occupying the necessary space.

*Flush Deck*—Solar panels would either have to be removable, mounted on the stern rail, or the special marine-grade ones that are designed to be walked on.

## ANCILLARY EQUIPMENT

In choosing an alternate energy system, one easily overlooked detail is the position of extraneous equipment on board your boat.

*Biminis*—We learned the hard way that Biminis must be taken into consideration when mounting solar panels. We originally secured our first 10W panel to the center of the aft cabin roof. The sun exposure was ideal until we left cool Cape Cod for hot Florida. Converting our rain tent into a Bimini, we rolled it out only to discover it effectively blocked any sun from reaching the panel. In fact, the panel was probably occupying the coolest spot on the boat.

In the tropics, boats with Biminis that extend the entire length of the cabin top have to mount solar panels on the stern, or keep them free to be moved around at will.

*Dinghies*—Dinghy stowage is often a problem. Stowed on your cabin top, it might cover a deck-mounted solar panel. In davits, a dinghy can prohibit stern gimbal-mounted panels (but not stern rail-mounted ones) and might interfere with either a pole-mounted wind unit or the handling of a water generator.

*Roller-Furling Jibs*—On boats with a roller-furling jib, a second headstay needs to be installed for hoisting a rigging-suspended wind generator. Roller-furling might also create some slight turbulence, although probably not enough to worry about, since we've seen many boats operating wind units with this arrangement.

*Equipment Off the Stern*—Before deciding on a water generator, consider what you already have hanging off the stern. Transom-hung rudders, outboard engines, self-steering units, taffrail logs, fishing lines, davits and boomkins will all effect the position or possibility of a water unit. Once again, we know from experience. Even on our wide boat we managed to mount the generator bracket of our water unit in exactly the wrong place on our first attempt. It's important to remember (as we didn't) that the trailing prop "walks" to port, covering about a 5° angle. While this is no problem on a straight transom with a rudder hung below the waterline (with the generator mounted to the extreme port side), our catamaran has two transom-hung rudders. After mounting the bracket and admiring our handiwork we realized the braided line would walk right into the adjacent (port) rudder.

*Passengers*—Don't mount your solar panel or pole-mounted wind generator in your favorite reading or sunning spot. The poor placement of our original 10W solar panel (remember the Bimini?) revealed it was occupying the children's favorite perch. While that particular panel was marine grade and impervious to small bottoms, it certainly wasn't getting much sun.

*Speed*—In the case of water generators, the boat speed is an important consideration. Our water unit, for instance, generates only ½–3 amps in the 3–4 knot cruising range but

*Selecting the Right Alternate Energy System*

11–15 amps in the 8–10 knot range. So, if you own a racy design or fast multihull, your water generator output will be very high and only occasional use will be necessary in most cases. Hamilton Ferris designed his Diving Plane with fast boats in mind. This device counters the tendency for the propeller to skip out of the water at high boat speeds (or at normal cruising speeds in rough weather). A slow 3-knot cruising boat may not be the best choice for a water unit, unless you mount an auxiliary propeller and gear it up to provide the high RPM necessary for high output. The additional drag of this drive system, however, would tend to further reduce your cruising speed.

## BOAT IMAGE

Many people complain that alternate energy systems don't look good. We've heard time and again how a solar system ruined the looks of a particular house. In our solar design business, we are continually striving to make solar energy features fit the look of a traditional New England home. Similarly, alternate energy systems on board a boat need not destroy the look of a particular craft.

If you have a wooden boat with classic lines, you might not want a pole-mounted or permanent mast-mounted wind unit. Nor would you perhaps choose a bright, shiny solar panel. But a panel with teak trim will match the skylights and blend nicely with your boat's traditional appearance.

In the same way, a sleek, modern sailboat might choose a sleek, modern-looking system. The Newick trimaran mentioned in the Multihulls section looked more like a rocket ship than a boat, with its matching solar panels mounted vertically on the reverse transom.

Whatever your boat's design, an alternate energy system can be found that fits comfortably on board, provides you with the necessary electrical power and enhances the appearance of your boat.

# CHAPTER 6

# BOAT USE AND HOW IT AFFECTS YOUR DECISION

The way you use your boat needs to be taken into account before purchasing an alternate energy system. The following boat-use categories will guide you in assessing your alternate energy system needs.

## THE WEEKEND SAILOR

A weekend sailor needs a system that can be left unattended yet will keep the battery topped up and ready for immediate use.

*Solar*—A small 7–10W solar panel will maintain the battery(s) by charging during the week while the boat is not in use. With care, it will handle an average load, including engine starter, cabin lights, a few instruments, and a few hours use of running lights. You might find you need the output of a 20W panel, in which case a second 10W panel can always be added later. However, purchasing a 20W panel at first usually is less expensive than two 10W panels.

## THE OCCASIONAL-CRUISE SAILOR

Once again you need a self-tending system, capable of a higher output for those occasional cruises.

*Solar*—Two or more 10W panels will cover the additional load of some longer cruises. A 20W, 30W, or 35W panel will also be a good choice, as the large panels are less expensive per watt up to 35W. The 35W panel is the best buy in non-marine panels as it is becoming the solar industry's standard size. Care has to be taken that the panels don't overcharge the battery when the boat is not in use for long periods of time, either by employing a voltage regulator, disconnecting one panel during the week (if two small ones are used), or by regularly checking the battery's state of charge.

*Wind*—A small permanent or pole-mount wind unit, like the Ampair, LVM or Thermax, can be left to charge batteries while you are off the boat. You may need a voltage regulator if you use the boat very infrequently. A large pole-mounted wind machine can be used while you are on board and tied off during your absences. But it might provide more electricity than you need. [If a bilge pump, light or other appliance is left in operation, a small (10W) solar panel may be needed to keep the battery up when the boat is unattended.] A rigging-suspended mount is the least practical system as it must continually be set up and taken down during your short times on the boat. To make matters worse, the wind often dies down at night, just when you need the electricity.

*Solar/Water*—A good hybrid combination is a 7–10W solar panel maintaining battery charge for weekend use and a water generator supplementing this output during cruises. The water unit can easily handle the larger load imposed by a cruise and is capable of generating anytime while you are underway.

## THE VOYAGING/LIVE-ABOARD SAILOR

Voyaging sailors or live-aboards will establish a routine similar to any home, resulting in a fairly constant daily electrical load. Any one of the AE systems, or a combination of systems will work well. If the boat is your primary residence for tax purposes, you will be eligible for federal energy tax credits (until 1986 as it now stands) that cover up to 40% of the cost of your system, as well as for state energy credits, where applicable.

*Staying In Harbor*—Your best choice here is solar system, a wind charger, or both. One or two 30–35W solar panels will handle your daily load, with the exception of refrigeration. Our light load is covered fall to spring in southern Florida by one 35W panel. Occasionally we need an extra boost, particularly when sailing all night with running lights. A large-prop wind unit is the better choice for a boat with a heavy electrical load. A water unit is a practical supplement during long sails, when running lights and electronic equipment can impose a big additional demand.

*At Dock*—If you plan to spend most of your time at a dock, a small mast-mounted wind generator or solar panel is your best option. Solar panels are not affected by dockside congestion and can be moved around the boat to avoid shading. You might want to take some precautions against theft, if that is a problem in your area. A wind unit can be mounted high enough to escape any interference from other boats and set on a platform to pivot with wind shifts.

*Harbor or Island-Hopping*—A large-prop wind unit is easily stowed for the major crossings, then resurrected on its pole mount for cruising the islands. All the other systems are fine for this type of cruising, or for coastal cruising from harbor to harbor.

## THE WORKING LIVE-ABOARD SAILOR

If you are away at work all day but home on the boat in the evenings, you need a system that is self-tending and capable of handling a live-aboard load.

*Solar*—You can mount a panel on each side of the deck, thus ensuring at least some solar exposure during the day. Or, set up one (or more) moveable panels each day before leaving for work. We have working friends who leave their 10W panel strapped to the mast

**Fig. 6–A.** Very likely the largest photovoltaic installation on a pleasure craft (1,060W peak). The possibilities for a charter vessel are unlimited (courtesy of A. E. G. Telefunken).

facing south for the best overall daily exposure to the sun.

*Wind*—From our experience, large-prop wind units *without* a reliable governor may not be a good choice for the working live-aboard sailor. Few of the large-prop generators are made for winds over 25-30 knots. We've seen units fly apart when a squall blew through an anchorage. Many units have overspeed control options. Even though well-made units will probably remain intact, high winds can damage the generator itself. They can be tied off while you are working, then restarted while you are on board. But this daily shutdown should be considered when sizing your system. The small-prop pole or permanent-mount units are fine for small to medium loads, although their output is minimal unless you live in an anchorage or marina with good wind exposure.

## THE OCEAN-CROSSING SAILOR

Solar panels, water generators and small-prop permanent or pole-mount British wind units are all excellent choices for the offshore sailor. The self-tending ability and fairly compact size of the Thermax unit also makes it suitable for passage-making, provided it is mounted on a sturdy, well-stayed pole. Water

*Boat Use and How It Afects Your Decision*

units can always be used underway and are easily capable of handling the extra load of running lights, autopilot and instruments. A small fixed or pole-mount wind unit would work well, although downwind its output is minimal, and a fixed-mount unit will only operate when the wind is forward of the beam. When sailing downwind, the boat's sails are a much better collector of the wind's energy. Take advantage of this energy surplus by towing a water generator. A large-diameter prop wind unit should be stowed during long sails in open water. The best choice for a boat undertaking long passages is a combination of systems. A water generator or solar panel can take over for permanently mounted wind units when sailing downwind. Also, most boats that undertake ocean passages still spend lengthy periods at anchor. A water generator with a wind conversion kit guarantees electrical generation both at sea and at anchor, as does a water-and-solar combination.

## SAILING WITH CHILDREN, PETS AND FRIENDS

With five-year-old twins on board *Kjersti*, we are aware just how careful you have to be with certain AE systems around children.

*Solar*—With solar panels, it's the panels themselves, not the people aboard, that are endangered. If space is limited and your panel is going to be mounted near walking feet, choose a marine-grade type. We had a low marine-grade Solarex panel on board until the children were four and had reached a new stage of responsibility. Now we have a 35W standard Solar Power Corp. panel, which should be left as undisturbed as possible, although even the tempered glass covers can take very rough treatment. With plenty of deck space on our catamaran, it's easy to keep the panel away from play activities.

*Wind*—The large-prop wind units potentially are the most dangerous systems aboard. If you choose a pole mount, it *must* be mounted high enough to be out of reach. Even a conscientious child can forget a spinning prop in the excitement of some game. Also, visitors—young and old—are faced with an additional hazard if they are unused to maneuvering around a wind generator. Rigging-suspended units also must be raised high enough to escape the reach of adults, let alone small children. When mounting a wind generator, remember that with children, even the unexpected can happen. We met one family whose children sailed their dinghy into their pole-mount wind generator, destroying the blade.

Pets also must be considered. Hamilton Ferris told us the story of a dog that momentarily got its hair caught in the spinning line of a water generator. Pets can also scratch solar panels with their claws.

## THE CHARTERBOAT OWNER

If your boat is used in bareboat chartering, you need a AE system that is undemanding, fairly foolproof, and capable of handling a large electrical load.

*Solar*—Permanently mounted, marine-grade solar panels (or standard panels if conditions permit) are an excellent choice. While probably not capable of handling the entire electrical load, they will save substantially on engine running time and thus are an excellent investment for charter companies and charter boat owners. One couple told us they were instructed to run the engine 3 hours every day; a realistic number for a boat with refrigeration.

*Wind*—A small-prop, permanently mounted wind unit has a potentially smaller output than other wind generators but is completely self-tending, out of the way, and able to generate electricity at anchor, at the dock or under sail.

*Water*—An auxiliary prop through the hull or an auxiliary generator on a free-wheeling transmission shaft is out of the way, self-tending and capable of a high output. A trailing-log water generator also is very little trouble and provides a high electrical output.

Alternate energy is ideal if you are operating a charter with captain and crew. The electrical demand is usually high, and so AE can make a significant contribution to lowering your overhead expenses. An AE system is a tax deductible business expense, and also eligible for federal energy tax credits if you use the boat as your primary residence. Even without tax credits, the equipment will quickly pay for itself in fuel savings alone. Charter guests also will appreciate not having to run the engine just to charge the batteries.

## THE RACING SAILOR

A racing yacht's alternate energy requirements include a small to medium electrical load imposed by electronic equipment during the race, sporadic demand—since the boat often is left unattended—limited stowage space and an aversion to increasing either weight or windage.

*Solar*—Our first choice are solar panels mounted either on the deck or stern rail. On deck, marine-grade panels can withstand being walked on by crew members. A textured surface, although not essential, will ensure secure footing, even when wet. Kyocera offers an optional plastic protective cover for its panels. Place these, or something similar, over the panels during times of heavy crew activity, or throughout short races. When the boat was not in use, the panels will keep the battery fully charged.

*Wind*—Large-diameter prop wind machines are not appropriate for racers. They require too much attention and, in the case of the rigging-suspended type, are only usable at anchor. Even small-prop permanent mast mounts create too much windage for serious racers, although the windage of a small-prop pole mount unit on the stern is minimal.

*Water*—Although the drag of a water generator is slight, it does increase with the square of the boat's velocity, making water generators most practical only for long-distance racers.

*Boat Use and How It Afects Your Decision*

## THE OFFSHORE RACING SAILOR

Numerous boats have already demonstrated how applicable alternate energy is to offshore racing. Unlike short-course racing, where small wind shifts and slight course adjustments are critical, the overall speed of an offshore racing boat is less affected by the drag, windage or weight of an AE system. Because ocean racing is demanding, any system should be self-tending. It also needs to be reliable and able to handle the heavy load of instruments, lights, radio, running lights, etc. Optimal is a combination system, providing high output and insurance in case of a breakdown.

*Solar*—This is the favorite method of generating electricity among ocean racers, although the cost/output ratio of solar panels is high. Use marine-grade panels, to withstand heavy-footed crewmen and the pounding of a hard-driven yacht in ocean seas. Solar will provide a near-constant source of charge. The number of panels used is dependent on space, electrical load, and their relationship to whatever other system is employed. The yacht *Fantasy* demonstrated the reliability of solar panels on the 1982 BOC single-handed around the world race. The skipper was using a four-panel Solec Solarcharger system to operate communication system, interior and navigation lights, instrumentation and autopilot. Even after one panel was bent by a large wave in a storm to a 90° angle, it continued to function for the remainder of the race.

*Water*—An auxiliary prop through the hull is a good choice, especially for shorthanded crews. There is nothing to set out or bring in as with the trailing-log type and nothing for large fish to nibble on. Ease of operation, however, must be weighed against adding another through-hull fitting on your boat. A generator like the Sail Charger, mounted on a freewheeling transmission shaft, is also a good choice, although many racing boats use feathering propellers. Of the trailing-log type generators, the Power-Log with no spinning line is probably the easiest to use and, with no internal electrical contacts, one of the most reliable. This unit made an impressive showing in the recent singlehanded around the world race on Richard Konkolski's *Nike III*. This isn't to say that other trailing-log types haven't performed admirably during offshore races, and most can be fitted for just over half the cost of the Power-Log. Two other water units that are good choices for racers are the extremely high output Silent-power water generator (higher output means less time you need to trail it) and the Diving Plane Hamilton Ferris unit, which keeps the propeller assembly submerged at high hull speeds and during rough weather.

## MAKING A DEADLINE

Inevitably, there are times when you are hard-pressed to keep to your schedule and the idea of doing anything to slow your boat even a fraction of a knot is unappealing. Even the best plans can involve meeting a tight deadline, like running an inlet on the proper tide or reaching an unlit destination before dark. It takes a great deal of inner fortitude to then toss your trailing-log water generator into the water, no matter how little drag the manufacturer claims it has. If you rely heavily on your water unit, allow for the slight reduction in speed when planning your trip. If you are still struggling to meet a deadline, practice a little energy conservation or do without electricity entirely. The preferred option is to slightly oversize your other AE system(s) to carry you through times when you might not want to tow a water generator, even for a few hours.

# CHAPTER 7

# THE IMPORTANCE OF CLIMATE

As important as any other alternate energy selection consideration, is where you sail. You have to consider not only the general climate, but local sailing areas, including harbors and even where you moor or dock. A wind generator that is excellent in the tradewinds might work poorly in the light summer winds of Long Island Sound. A water unit that works well offshore might be impossible in lobster pot-infested Maine waters. If you are planning a trip around the world you will need a system that can adapt to a variety of climates. Whatever system you choose and however you mount it on your boat, it must be capable of operating well wherever you sail and live.

## SOLAR RADIATION

Obviously, solar energy devices work best in sunny climates. However, most areas have sufficient sunshine to justify a solar system. In a solar panel, as long as there is light some charge is going into the battery, even on a hazy day. Depending on the number of cells, most solar panels will start charging at a sunlight level equivalent to an overcast day (10 mW/cm).

### Available Solar Radiation for the Continental U.S.

- The southwest U.S. from western tip of Texas to Yuma, Arizona has the most available sunlight in the continental U.S. Unfortunately, it's not the best sailing area.
- The southern coast of California, west Texas and southern Florida have very high sunlight levels throughout the year.
- Central California, east Texas and the Gulf states, central and northern Florida, Georgia and South Carolina are year-round cruising areas with an abundant level of sunlight throughout the year.
- The East Coast from Maine to North Carolina and the Great Lakes (except Ontario) have good sunlight levels from early spring through mid-fall.
- Lake Ontario and St. Lawrence River offer fair amounts of sunlight during the sailing season.

### Solar Radiation Outside the Continental U.S.

Everywhere south of latitude 28N, including the Bahamas, Puerto Rico, the Caribbean and Mexico, is excellent for generating solar energy. The Hawaiian Islands also have very good year-round sunlight levels. The coast of British Columbia and southeast Alaska have poor amounts of sunshine. During our summers spent in and around Juneau, one of the few days the sun came out all the stores

FEBRUARY

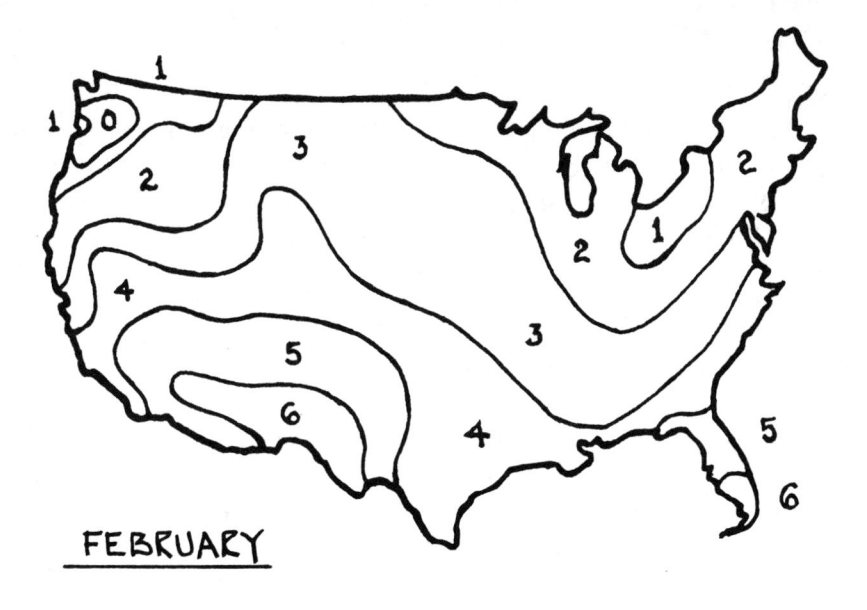

**Fig. 7-A.** *Mean Daily Solar Radiation*
*Legend* (In Langleys; one Langley = 3.69 BTU/ft$^2$)

| | |
|---|---|
| 0 — <150 | 6 — 400–450 |
| 1 — 150–200 | 7 — 450–500 |
| 2 — 200–250 | 8 — 500–550 |
| 3 — 250–300 | 9 — 550–600 |
| 4 — 300–350 | 10 — 600–650 |
| 5 — 350–400 | 11 — >650 |

Note: Information provided by the Environmental Science Services Administration.

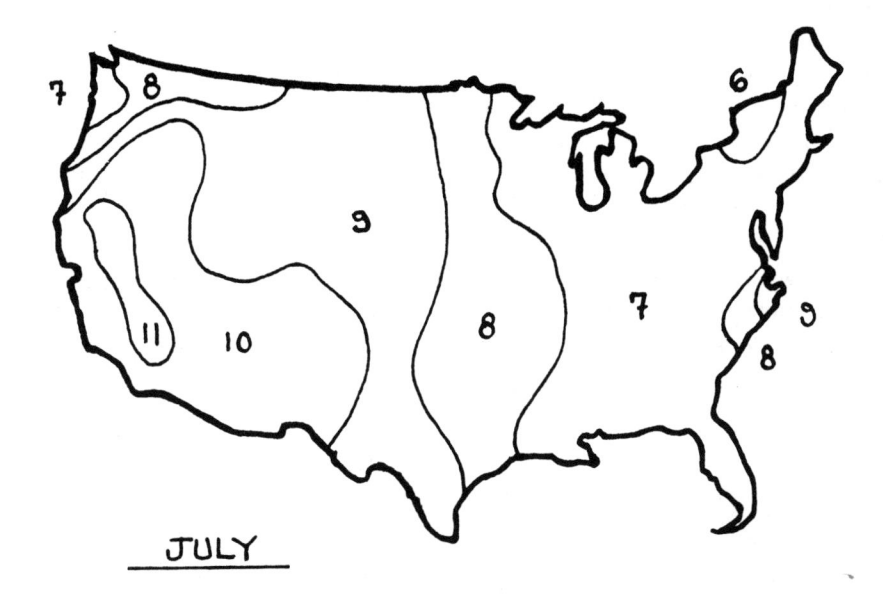

JULY

closed and they declared a holiday. In general, from 40°N latitude to 40°S latitude around the entire world there is adequate sunshine to easily support the use of photovoltaic energy sources.

Anyone voyaging extensively north to south will find that sunlight levels will even out in the course of a passage, and thus will find photovoltaics a good overall choice of system. Sailors who expect to put many miles under the keel should select a hybrid system, with solar as one of the alternate energy-generating systems.

## Micro Climate

Micro-climate refers to localized conditions that occur within a larger climate, often caused by topographical characteristics or differences in water temperature. Cape Cod, for instance, experiences frequent fogs and strong onshore winds on its south shore during the summer, while there is little fog and light winds to the north. Maine's coastal islands are often constantly enveloped in fog that burns off quickly along the mainland coast. Cape Hatteras on the east coast and Mendocino on the west are notorious for attracting bad weather, which often does not affect the surrounding areas.

Before choosing a solar system, study the micro-climates in your area and, if necessary, oversize your system slightly to carry you through those times of little sunshine. Don't be misled by an apparent lack of sun in the areas that rate lower on the sunshine scale. Northern areas, such as New England or the Pacific Northwest, may be only fair for solar in the winter but quite adequate during the summer sailing months. Even hazy skies will give close to maximum output. As testimony to this, relatives of ours in central New Hampshire generate all their home electricity for lights, radio, TV, and tape deck with two 35W solar panels. Similarly, on Monhegan Island off the coast of Maine (another area not known for its sunshine) photovoltaic solar panels provide a large part of many of the residents' electrical demand.

Micro-climate will not greatly effect your solar output on an extended voyage, particularly if you carry a backup system to supplement areas and times with poorer sunshine.

## Mini-Micro Climate

This refers to the immediate climate of your dock, anchorage, mooring or sailing area. A greater abundance of sunshine in one part of a neighborhood is a good example of a mini-micro climate. Some states now protect solar-heated buildings from having their sun blocked by new construction.

*Dock*—In an extreme case, your sunlight might be blocked in a slip by two large yachts docked on either side of you. A disruption of your mini-micro climate might be an issue to take up with your marina dockmaster. Large buildings, high docks or poles all are potential, but not insurmountable, problems. You can usually, after all, find a better place to dock.

*Anchor/Mooring*—Here again, buildings, trees, high ground and possibly other boats can interfere with sunlight. It's an important issue to consider if you plan on remaining in one place for a long time. But lying to an achor or mooring most often ensures mobility, and thus the ability to avoid any debilitating sun blockage.

## PREVAILING WINDS

While these two factors appear to have little effect on solar energy generation, they do influence your boat's position at anchor and therefore its exposure to sunlight. For instance, on the south side of Cape Cod, the prevailing winds are southwest during the summer season. As any boat anchored or moored will naturally head SSW, solar panels preferably are mounted tilted slightly towards the front of the boat, horizontal on the foredeck, or placed one on each side to catch sun throughout the day. Similarly, if you consistently anchor or moor in one place affected by a strong current, mount your panel where it will face the sun for the longest possible time. Or, leave it movable.

Throughout the world, certain areas experience stronger, more constant winds than others. Regions exposed to tradewinds or westerlies, for instance, are going to be breezier than those near the doldrums, or those places closer to the Equator. The book "Ocean Passages of the World" provides an excellent source of the world's prevailing wind strengths and directions, as do Pilot Charts.

### Coastal Winds for U.S.

As a general rule of thumb, all U.S. coastal areas experience good average windspeeds. The lack of obstructions on most coasts guarantees sufficient exposure to the prevailing westerlies that influence most of the U.S., except for the east coast of Florida and the Gulf of Mexico, which experience predominantly easterly winds throughout the year. Another rule of thumb is that average summer windspeeds are lower than winter ones.

- Maine to North Carolina experiences good year-round average windspeed. The south coast of Cape Cod to Block Island is an exceptionally windy area during the cruising season. South Carolina and Georgia are moderately windy during the cruising season. The east coast of Florida and Gulf are good and southern Florida, the Keys, and S.E. Texas are very windy regions.
- The U.S. West Coast offers good exposure to westerlies, with the wind generally modulating the further south one goes. Mexico is fair while southern California offers good average windspeed. The central California coast to the Canadian border is especially good during the cruising season. The San Francisco area and Washington State are regions of very high average windspeeds.

### Micro Climate

Local wind conditions are influenced by prevailing winds, local temperature imbalance and geography. Areas of slower winds typically are landlocked areas that are blocked from prevailing winds. Long Island Sound, the Chesapeake and the Intracoastal Waterway are all well known cruising areas where

*The Importance of Climate*

the adjacent land blocks the lighter summer winds. Summer sea breezes, occurring very near the coast, often build in intensity during the day, but die down at night. The position of land in a given area and its effect on the wind will create specific wind micro-climates. On Cape Cod, for example, the north shore is blocked from the southwest and thus experiences light winds most of the summer. Yet on the south shore, the exposure to the prevailing southwesterlies (often enhanced by sea breezes) is excellent. Not only is the sailing south of Cape Cod excellent, but it is an ideal place to use a wind generator to make electricity. Another good example is Puerto Rico, where the north coast is exposed to the nearly constant tradewinds, while on the south shore the average wind velocity is much lower. In the Bahamas, the position of the islands has created various micro-climates. The average wind velocity increases towards the southern half, which lies more directly in the path of the tradewinds. The northern islands experience winds similar to the east coast of Florida. But, because they are low-lying and located far out in the ocean, their average wind strength is greater than Florida's. By familiarizing yourself with the prevailing local winds and their average strength and by considering what effect any nearby land mass will have on them, you can determine whether you are exposed to enough usable wind to make a wind unit feasible on your boat.

## Mini-Micro Climate

Microcosmic anomalies are particularly important to consider when deciding whether to acquire a wind generator. High land, currents, the position of the shore—even the size and number of nearby trees—all effect the amount of wind that reaches a given spot. While anchored in Vero Beach, Florida, we discovered that a tall group of pines 300 yards across the harbor to the southeast created a turbulent effect on our wind generator when the wind blew from that direction. Only *Kjersti* was effected. A boat anchored less than 100 yards away had a wind unit that completely escaped the disturbed wind effect. It was fascinating to see what a powerful effect even a distant stand of trees had on our wind generator. The situation was reversed when the wind veered to the southwest. A nearby island prevented the wind from reaching the other boat while we continued to generate.

If you are a sailor who prefers only the snuggest of anchorages, a wind unit may not be the right choice for you. You must have good exposure to the wind for the unit to operate efficiently. This does not mean you must seek anchorages exposed to the sea. If your load is small and the output from your wind generator is high, it will only require periodic operation, something to consider when sizing your system. But if you intend to stay somewhere for a length of time and are dependent on the wind to supply your electrical needs, you will have to choose a spot where the wind blows freely—or have another means of generating electricity. We spent one winter in Florida anchored close to an island for protection from the winter winds. Most of the time our wind generator was not operating, but our large solar panel provided us with sufficient electricity. Other-

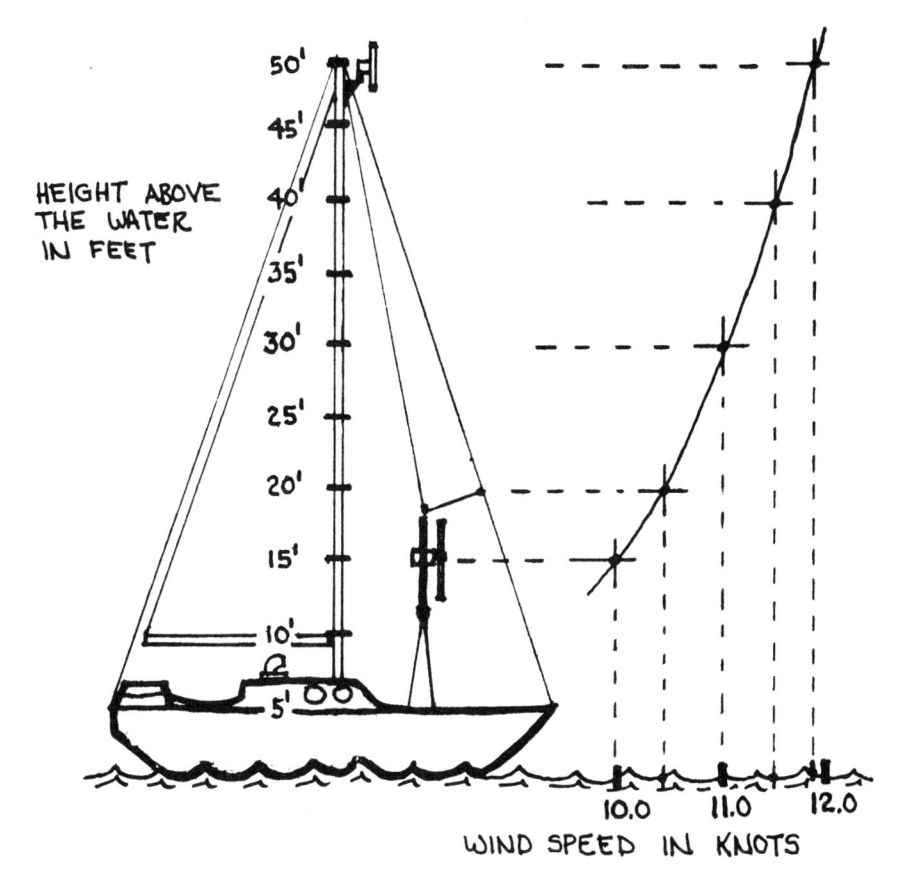

**Fig. 7-B.** An example of windspeed as a function of height above the water.

wise, we would have been forced to anchor in a more exposed, uncomfortable spot.

When anchored in a tidal current that determines your boat's heading, make sure your wind unit can pivot to follow the wind. The height of the wind generator off the water's surface will effect its overall output. Even in well exposed areas, wind speed increases with height. This is due to the boundary layer effect shown in Fig. 7-B. This makes a significant difference, especially when you consider that the power output varies with the cube of the windspeed.

The windspeed at spreader height can be 1.25 times the windspeed at deck level, but that difference doubles the output of a wind generator. This rule should be taken into account when you examine the output of small units, such as the Ampair, that can be mounted on the upper part of the mast.

# WATER

Geographic area can effect a water generator two ways. Some areas, like the Bahamas and Florida Keys, have extremely shallow waters

*The Importance of Climate*

that warrant some caution when using a trailing log-type water generator. Areas with a abundance of seaweed or fishing paraphernalia also will pose a problem. The lobster pots of Maine or crab pots of the Chesapeake might prevent use of a water generator when it is needed. In these areas, an auxiliary water propeller and generator or a generator on a freewheeling transmission shaft is preferable. On the other hand, a trailing log unit can be retrieved while sailing over pot-infested waters, while an auxiliary propeller gives pot lines two targets to foul. A means of locking the prop shaft will cut down on the risk of snagging things in the water and eliminate unnecessary wear on shaft and transmission when the batteries are charged.

The local climate also can effect a water generator indirectly by creating either calms that give you little or no sail power or high winds, which create boat speeds excessive enough to require stowing the unit.

# CHAPTER 8

# HYBRID SYSTEMS AND OTHER ALTERNATIVES

Independence on a boat means being prepared. If you carry spare ground tackle or sails, you become less likely to be caught short in an emergency or dependent on finding yachting facilities. The same rule applies to alternate energy systems. Carrying more than one AE system, or a hybrid system, increases your ability to generate in all conditions, making you more self-sufficient.

A hybrid system is most practical for three types of sailors:

- Those who intend to sail far from home
- Those with large electrical loads
- Those who strive to be as completely independent as possible.

Despite the advantages of alternate energy, it can be frustrating at times. People with wind generators suffer through days with no wind and others with too much. People dependent on solar panels still use electricity, even on cloudy days. Or, they find themselves sailing with the sails shading the panels. Those using a water generator can be becalmed or find themselves moving too fast.

But it's rare when conditions are not favorable for at least *one* of these systems to be generating. With a hybrid system, you increase your electrical output while protecting yourself against breakdown. Let's take a look at various combinations for a variety of possible situations.

## A WIND/WATER HYBRID SYSTEM

Several manufacturers of boat AE systems, including Ampair Products, Everfair Enterprises, Friendly Energy, Hamilton Ferris, Redwing, W.A.S.P., and Wind Turbine Industries, sell hybrid kits that convert into either wind or water electrical generating systems.

These two systems complement each other nicely, as water units perform excellently while underway and wind units are best used at anchor. With a permanent or pole-mount wind unit you have the option of using both systems while underway (if you have two separate generators), although some points of sail are less favorable for wind-driven generators. For boats with lighter electrical loads, a rigging-suspended mount is practical, as it eliminates the windage of a pole or permanent-mount unit, generates electricity at anchor and stows when the water unit is in use. A large propeller pole-mount should not be set up while sailing where there is a chance of big seas. Even on a sturdy, well supported pole, the extra 20–30 lbs. aloft are a potential hazard when the boat is rolling heavily. The generator and prop can be removed from the pole during a long passage or in bad conditions and remounted later. You might try jury-rigging a wind/water hybrid system yourself, using the generator from your wind unit on a trailing water prop (see Chapter 11).

A wind/water hybrid system is equally appropriate for long passage-makers and gunkholers. The voyaging sailor can use the water unit during the passage, then set up the wind unit for those revitalizing long stays at anchor. The water generator enables the coastal cruiser to keep the battery topped off during short hops, thus reducing the need each night to raise the wind unit (with a rigging-suspended type) or to seek exposed anchorages.

## A WATER/SOLAR HYBRID SYSTEM

The combination of a water/solar system is practical for sailors with heavy loads and for those with a light load who periodically require extra power for running lights and cabin lights during night sails or offshore passages. It's also an excellent choice for sailors who like ease of operation and the independence of two reliable systems.

A water generator potentially can produce 100 amps or more a day, making it an ideal unit for boats with a heavy load or those undertaking frequent or lengthy sails. A high output water unit can, in good conditions, handle the heavy drain of an SSB transmitting radio, refrigeration or autopilot while solar maintains the basic lighting load.

For the sailor with a light electrical load needing an additional occasional boost to power running lights, etc., the water generator guarantees electrical output when underway. We met a sailor who had just circumnavigated the world using a 35W solar panel as his primary source of electricity. He told us the panel did a nice job meeting the regular load of lights, radio and instruments, but couldn't meet the demand of bright (therefore power-consuming) running lights. He had no deck space for additional solar panels and with both fishing lines and a taffrail log trailing off the stern of his double-ended Tahiti ketch, felt he couldn't use a water generator. We had three suggestions for him:

- Mount one or two solar panels on the stern rail.
- Put a water generator on a free-wheeling prop shaft off the engine transmission, if possible.
- Trail a water generator from a small outrigger to port and the taffrail log on a similar arrangement to starboard. As the generator output is high and his load is fairly low, 3–5 hours of operation a day will meet his running light load, leaving the rest of the time available for trailing fishing lines. The water generator also can boost the electrical supply on cloudy days when the solar panel is putting out little energy.

If you have a boat with a very light to light load, a 35W solar panel will take care of all your basic necessities. Adding a water generator is an easy, inexpensive way to increase your options. This is basically how we operate while traveling. Our solar panel keeps the battery topped off in most conditions. On long or night sails we trail the water unit to give the battery an extra boost. This hybrid system allows us to indulge ourselves without worrying about draining the battery.

## A WIND/SOLAR HYBRID SYSTEM

When cruising with nightly stops, it can get tedious every evening to raise a rigging-suspended wind unit. Often we don't feel like raising the generator at the end of the day or find ourselves anchored in a spot with little wind. By adding a 9W solar panel we freed ourselves from the nightly chore. While the small panel doesn't meet our electrical load, it means we only have to raise the wind unit every 3–4 days to charge the battery. When we settle for a few months in one spot, both the solar panel and wind generator are left in place to generate. Recently we traded our 9W panel for a 35W unit that does a far better job of meeting our total electrical demand. Now our wind or water backup unit is needed only occasionally.

While pole and permanent-mount wind units are always in position and ready to generate, boats with these systems inevitably encounter anchorages with minimal winds or downwind passages when a wind generator is not effective. A solar panel can provide the necessary backup with no extra work. In a snug harbor or an idyllic, unexposed anchorage, the solar panel provides the option of remaining in these places while continuing to generate electricity.

## A SOLAR/WIND/WATER HYBRID SYSTEM

With the exception of boats with very heavy loads, carrying all three AE systems guarantees electrical self-sufficiency and provides the security of knowing you will always be able to generate electricity without resorting to the engine.

If you would like a maxi-AE system, get a wind/water combination unit from one manufacturer and the largest (highest output) solar panel you can fit on board. If the panel can be placed away from traffic, shop around for the least expensive standard unit. A wind/water system plus a standard 30–40W panel will cost $1000–$1400, more if you buy

**Fig. 8–A.** Our hybrid alternate energy system aboard *Kjersti* . . . A 35W solar panel, and a wind generator that converts to a trailing log water generator.

a marine-grade panel. But this is a modest price for such reliable, long-lasting equipment when compared to other cruising expenses.

## SUPPLEMENTING ALTERNATE ENERGY SYSTEMS

### Portable Generators

Portable generators run efficiently on gasoline, are easy to stow, lightweight and relatively inexpensive to repair. In addition, they can supply 110V AC power for tools or appliances, as well as 12V DC for charging batteries. The following is a partial list of available portable generators and their specifications:

These generators should be used as a backup to an AE system, or to supply 110V AC power. Relying on one as your primary generating source has the following problems:

- They all use explosive gasoline as fuel, with all its inherent polluting characteristics. If you have a diesel marine engine and no other reason to carry gasoline other than to power a portable generator, an AE system is all the more practical.
- It relies on a higher technology (gasoline engine) to provide something easily obtained from a much simpler alternate energy generator. In the Bahamas we met sailors on two boats who were having mechanical problems with their gen-

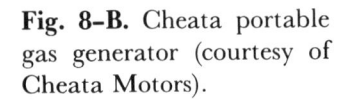

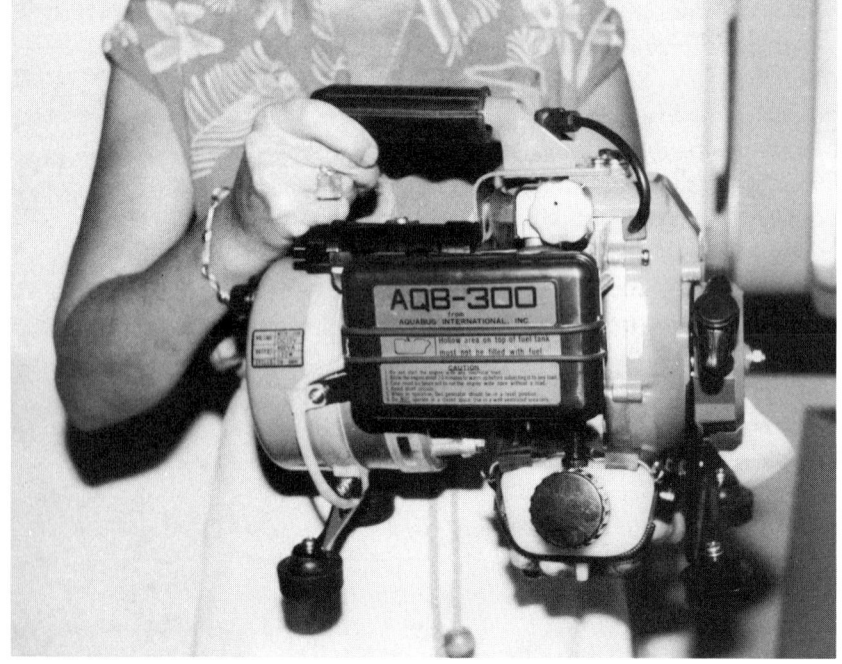

**Fig. 8-B.** Cheata portable gas generator (courtesy of Cheata Motors).

|  | Maximum Rated Output @ 12V DC | Current @ 110V AC | Size L x W x H | Weight | Fuel Capacity | Fuel Consumption | Manufacturer's Suggested Retail Price |
|---|---|---|---|---|---|---|---|
| **Cheata** | | | | | | | |
| • AQB=300 | 10.A | 2.8A | 11.81" x 8.31" x 11.3" | 18.7 lbs. | 0.15 gal. | 0.097 gal./hr. | $289 |
| • AQB=250 | 20.A | (none) | 11.81" x 8.31" x 11.3" | 18.7 lbs. | 0.15 gal. | .097 gal./hr. | $289 |
| **Honda** | | | | | | | |
| • EM500 | 8.3A | 4.1A | 14.0" x 9.8" x 12.8" | 40.0 | 0.5 gal. | 0.13 gal./hr. | $379 |
| • EM600 | 8.3A | 5.0A | 14.4" x 9.8" x 12.8" | 41.9 lbs. | 0.5 gal. | 0.13 gal./hr. | $389 |
| **Robin** | | | | | | | |
| • R600 | 8.3A | 5.45A | 14.6" x 10.4" x 13.6" | 40.7 lbs. | 0.53 gal. | 0.13 gal./hr. | $385 |
| • R1200 | 8.3A | 10.9A | 19.1" x 11.3" x 16.1" | 60.6 lbs. | 0.90 gal. | 0.23 gal./hr. | $575 |
| **Kawasaki** | | | | | | | |
| • KG550B | 8.5A | 5.0A | 15.1" x 9.5" x 14.3" | 38.6 lbs. | 0.73 gal. | 0.12 gal./hr. | $399 |
| • KG750B | 8.5A | 6.8A | 18.8" x 12.4" x 15.4" | 61.7 lbs. | 0.79 gal. | 0.18 gal./hr. | $479 |

erators. One boat was forced to use its less efficient marine outboard alternator while the other had an outboard with no alternator, and consequently no electricity.

- Despite what some people say, most portable generators are noisy, although they are relatively quiet compared to some marine engines.

You can avoid the whole 110 AC power requirement altogether by buying only 12V DC operated appliances. Many 12V appliances are available today, including fans, TVs, stereos, tape decks, radios, galley equipment and power tools. Available tools include drills (with attachments for sanding and buffing), soldering irons and sabre saws. Write for catalogs from:

Solarex/Energy Sciences
16728 Oakmont Avenue
Gaithersburg, MD 20877

**Fig. 8–C.** Robin portable gas generator (courtesy of Nagata American Corp.).

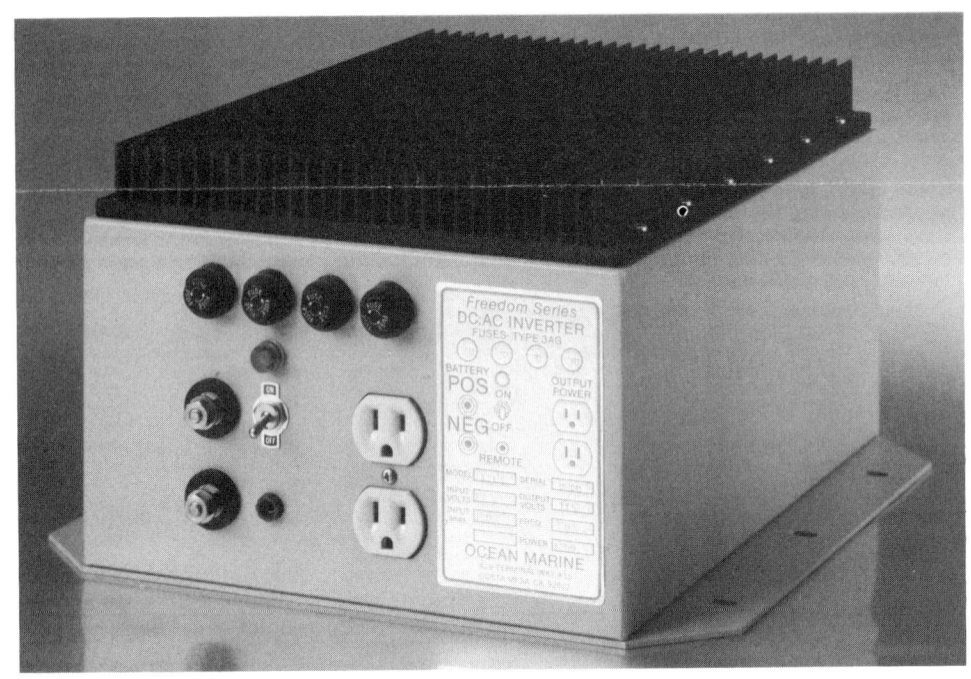

**Fig. 8-D.** A Freedom 12V DC to 110V AC inverter (courtesy of Freedom).

Imtra Corporation
151 Mystic Avenue
Medford, MA 02155

We were thrilled to find that some of the new electronic typewriters are powered by 6 or 9V DC. The Canon Telestar weighs only 4–5 lbs. and with a small converter can plug into your 12V battery. The best news is that it draws only ½ amp and costs about $225.

## Inverters

A second alternative is to get an inverter that changes 12V DC into 110V AC. These are widely available, their cost corresponding to their rated output or capacity. Since energy is changing form, electrical losses occur in the conversion. These losses begin the moment you turn the inverter on, even before you use an appliance (this is called the stand-by power drain). Inverters produce power that is fine for power tools and motors but may not operate (and can possibly damage) TVs, stereos and other sensitive electronic equipment. To power these appliances it is necessary to use an inverter with an output similar to household power. The inverter manufacturer will tell you what appliances its unit will operate. Accurate voltage regulation also is essential for operating TVs, computers, etc. You must find an inverter that keeps voltage constant regardless of battery voltage fluctuation.

There are two basic types of inverters. One

*Assessing Your Needs*

is a DC motor and an AC generator together in the same unit, often wound on the same shaft. This is called a motor generator or rotary inverter. Another type, the electronic inverter, uses solid-state circuitry to convert DC to AC. As a rule, rotary inverters are used for larger power tools and other motors with high starting torque. Electronic inverters are used for operating small tools and electronic equipment. An electronic inverter usually weighs less than the rotary type and has a much lower stand-by power drain, as low as 1–2% of rated power.

A few inverter manufacturers/suppliers are listed below:

### Electronic Inverters

Ocean Marine Instruments
629 Terminal Way #15
Costa Mesa, CA 92627

Heart Interface
1626 S. 341st Place
Federal Way, WA 98003

### Rotary Inverters

Winco Div. of Dyna Technology, Inc.
7850 Metro Parkway
Minneapolis, MN 55420

Thermax Corporation
One Mill Street
Burlington, VT 05401

## Boosting Engine Alternator Output

Most marine engines have some form of generator or alternator that serves as a logical backup to your AE system. You might as well be generating during those inevitable times

**Fig. 8-E.** A Spa Creek Instruments manual alternator control (courtesy of Spa Creek).

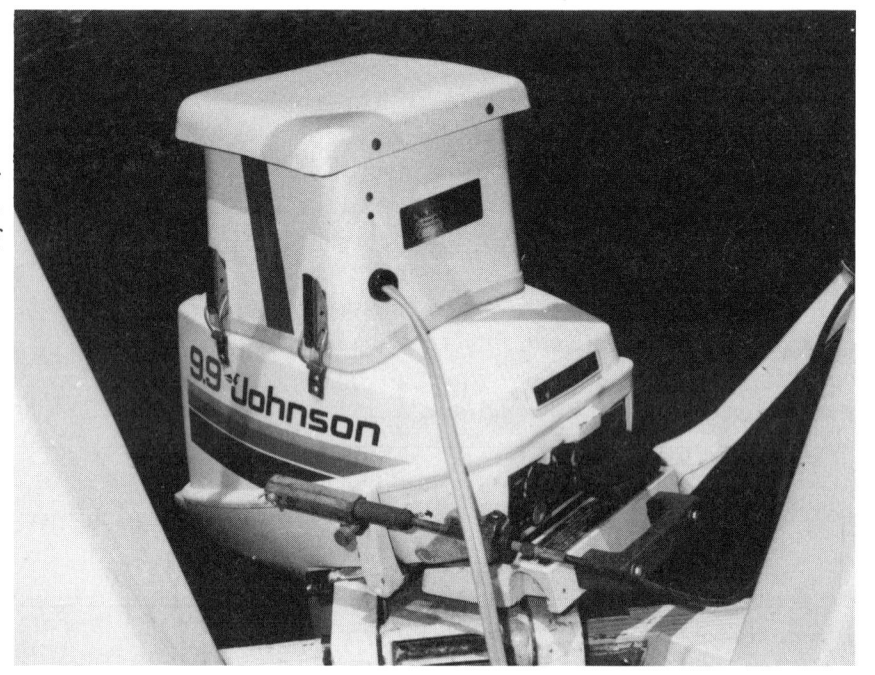

**Fig. 8-F.** Polder Power Pack mounted on an outboard motor (courtesy of Polder Marine Electric).

the engine is running. But when you do motor, your electrical output should be as high as possible.

Inboard alternators are voltage-regulated so that one rated at 35 amps is really only putting out that much current until the battery voltage begins to rise. The charge rate then will drop to 8–10 amps to keep the batteries from overcharging. Since most sailors run their engine for only short periods, the output could be much higher. Spa Creek Instruments offers a manual alternator output control and an automatic version, called the Automac. These devices allow you to manually regulate your alternator to keep it at, or near, its rated output. With the manual control, care must be taken to not overheat the alternator. This is monitored and controlled automatically by the Automac. You can get one of two models, depending on whether your voltage regulator is internal or external to your alternator.

If you are good with electrical things, "The 12 Volt Doctor's Handbook" describes how to make your own manual control and Spa Creek can supply the parts. Hamilton Y. Ferris II Co. offers a similar alternator control known as Dial-A-Charge.

Outboard engines used for sailboat auxiliaries have much smaller alternators, with 5–10 amp ratings. Like inboards, they do not sustain their rated output as battery voltage rises. There are several ways to boost the electrical output possible from an outboard motor. The most attractive is the Polder Power Pack, a device that attaches directly to the top of an outboard motor cover. The flywheel shaft must be accessible under this

*Assessing Your Needs*

cover as the Power Pack connects directly to it. Get specifications by writing to:

Polder Marine Electric, Inc.
5494 Cadbury Road
Whittier, CA 90601

Another method of boosting your outboard's changing current output is by dismantling and rewinding the alternator with heavier wire. This permits much higher outputs with the same alternator. This is beyond the capabilities of most sailors, but a trained armature serviceman can do it for you. We met a former mechanic on a 26′ Catalina who had changed the alternator output on his small Honda outboard from 5 to 20 amps. We will add, though, that while generating, he ran the engine at full throttle, nearly deafening everyone in the anchorage. This would have been an excellent backup to an AE system, allowing him to top up the battery whenever the boat was under power but avoiding the need to run the engine while at anchor.

*Hybrid Systems and Other Alternatives*

# PART III
# Selecting, Installing and Operating Your System

# CHAPTER 9

# MATCHING A SYSTEM TO YOUR NEEDS

Now that you're familiar with various alternate energy systems and have taken a close look at your own requirements, it's time to start putting together an AE system that's right for you and your boat. But before you start shopping around, there are a few practical considerations to remember in sizing your system, based on an estimate of usable daily output from various AE systems.

## NOTES ON SIZING YOUR ALTERNATE ENERGY SYSTEM

AE systems, on land as well as on boats, don't usually provide the instantaneous electrical output of gasoline and diesel-driven generators. Instead, they provide a fairly constant low to medium level of output. This means you must size your system so its average output, which we will estimate later in the chapter, is slightly higher than your average electrical load. When calculating the average output of a system, allowance must be made for times with less than optimum generating conditions. By sizing your system this way, you will avoid an electrical debt larger than your AE system can repay. A well-sized system for a particular load will always be building up reserve energy in the battery(s). Low-output generators such as solar panels, small-diameter propeller wind units and water generators on slow boats will take a few days to build up this reserve.

If, for any reason, you can only carry an AE system with an average output lower than your average consumption, you will periodically need some means to quickly boost your battery back to full charge. Two good solutions are either a manual control on your engine alternator (or a Polder Power Pack on your outboard engine) or a small portable generator (see Chapter 8). The manual alternator control or the Polder Power Pack will supply 35–50 amp-hours of electricity in one hour of generating. A standard 500W portable generator will provide 8–10 or more amps per hour in the DC battery charging mode. When used in conjunction with an AE system that is undersized for your load, these two options offer the most efficient way to pay off an electrical debt. Some homes switching to an alternate energy electrical supply operate the same way with a backup diesel generator. The AE system handles the constant light load (lighting, TV, radio, small appliances, etc.) while the diesel generator is only turned on occasionally to handle the heavy demand (water pump, washing machine, etc.).

Another way to deal with a periodic electrical debt is to manually or automatically disconnect your non-essential electrical load (everything except bilge pump, engine starter, running lights) when your batteries reach a certain level of discharge (see Load Disconnects in Chapter 10). This is an extremely simple and useful way to avoid building a large electrical debt. Most photovoltaic and wind generator systems for homes use this type of electrical disconnect. We used a manual version on *Kjersti* while going down the Intracoastal Waterway with only a 9W solar panel for generating electricity (no engine alternator). As the panel had much too low an output for our daily load, we only used electric lights and our radio on sunny days when the battery was fully charged. Otherwise, kerosene lights were put into use. While this system kept our battery from a deep discharge, we found it annoying not to be able to use electricity when we wanted, particularly in early evening, when the children played in their cabin. From experience we learned it's much easier to oversize an AE system than worry about having a constant supply of electricity.

Automatic load disconnects operate by sensing your battery's state of charge and completely cutting out the non-essential load (which is whatever you select) when the battery, or bank of batteries, gets below 50–60% of full charge. When the battery recovers, the device automatically reconnects the load.

Sizing an AE system is difficult because a number of factors determine your load, including the output of a particular system, storage capacity, and rate of consumption. While the factors in sizing a system can seem endless, one or two variables inevitably become the most important. Nine times out of ten someone sizing an AE home heating system will make detailed calculations only to discover the final decision is contingent on the extent of the home's south-facing exposure. The same holds true with boats. Review your personal situation, get a rough estimate of the output of the system that most appeals to you, and match the two with an average of your electrical load (including any appliances you plan to purchase in the near future). If you find your choice of system is slightly undersized, you can always add to it later (additional solar panels, a hybrid system, etc.). If your planned AE system is slightly oversized, enjoy the peace of mind and surplus of electricity. A surplus of electricity is easily put to use.

## ESTIMATING USABLE DAILY ALTERNATE ENERGY OUTPUT

In Chapter 4 we estimated the average daily electrical load of four typical boats. This is a reasonably accurate estimate since the day-to-day electrical consumption on a boat is quite predictable. For ease of comparison, we will estimate an AE system's average output on a daily basis, keeping in mind that it is subject to the vagaries of sunlight windspeed and boatspeed. You may generate lots of electricity one day and much less the next. Another thing to keep in mind is that manufacturers' rated output for equipment is usually for a partially discharged battery, and will be lower as the battery reaches full charge. As mentioned earlier, it would be nice to see wind and water generators rated by an independent testing facility in carefully controlled conditions, the way solar panels

*Selecting, Installing and Operating Your System*

## AVERAGE DAILY OUTPUT OF MARINE ALTERNATE ENERGY SYSTEMS
### SOLAR

| Panel Size | *Fair to Good**  *Sunshine Region*  *(Amp-Hrs./Day)* | *Very Good**  *to Excellent*  *Sunshine Region*  *(Amp-Hrs./Day)* |
|---|---|---|
| 10W | 2.5 | 3.5 |
| 20W | 5.0 | 7.0 |
| 30W | 8.0 | 11.0 |
| 35W | 9.0 | 12.5 |
| 40W | 10.0 | 14.0 |
| 44W | 11.0 | 15.5 |

* Above numbers assume clean panel cover, and panel is mounted on a horizontal surface with good exposure to the sun.

### WIND

| Model | *Light**  *Average Wind*  *(Amp-Hrs./Day)* | *Moderate**  *Average Wind*  *(Amp-Hrs./Day)* | *Strong**  *Average Wind*  *(Amp-Hrs./Day)* |
|---|---|---|---|
| LVM 25 | 1 | 4 | 10 |
| LVM 50 | 3 | 8 | 18 |
| Aquair 50 | 3 | 10 | 25 |
| Ampair 100 | 6 | 25 | 50 |
| Thermax "Windstream" | 6 | 25 | 50 |
| Seebreeze | 10 | 46 | 90 |
| Redwing | 10 | 50 | 100 |
| **Windcharger W200-12 | 8 | 50 | 110 |
| Hamilton Ferris II | 10 | 50 | 100 |
| W.A.S.P. | 18 | 50 | 100 |
| Friendly Energy | 18 | 50 | 100 |
| Windbugger | 18 | 65 | 130 |
| Everfair D54-2 | 18 | 65 | 130 |
| **Everfair D72-2 | 25 | 80 | 160 |
| **Wind Turbine (50 AMP) | 25 | 110 | 240 |

* Above numbers assume unobstructed exposure to the wind.
**These units employ a 72" diameter propeller.

*Matching a System to Your Needs*

## WATER

| Model | Boat Sailing At 4 Knots (Amp-Hrs./Day) | Boat Sailing At 5 Knots (Amp-Hrs./Day) | Boat Sailing At 6 Knots (Amp-Hrs./Day) |
|---|---|---|---|
| Aquair 50 | 20 | 50 | 65 |
| Aquair 100 | 40 | 80 | 120 |
| Redwing (water unit) | 40 | 80 | 120 |
| W.A.S.P. (sea generator) | 40 | 80 | 120 |
| Hamilton Ferris II (water trolling unit) | 50 | 110 | 170 |
| Everfair Enterprises | | | |
|   WG-1 | 65 | 125 | 175 |
|   PSD-1 | Depends on boat's prop size and configuration. | | |
| Power Log | 35 | 130 | 170 |
| Wind Turbine Ind. (water trolling unit) | 65 | 160 | 250 |
| Sail Charger | Depends on boat's prop size and configuration. | | |

have been rated. In any case, before you purchase a system, make sure the manufacturer will guarantee his listed output.

## FOUR TYPICAL ALTERNATE ENERGY SYSTEM INSTALLATIONS

Using the AE systems electrical output chart and the electrical load of four hypothetical cruising boats in Chapter 4, we can see what systems will satisfy that electrical load. The numbers on the output chart assume good solar and wind exposure. If you anchor in harbors well protected from the wind or can't move your solar panel around to avoid shadows, some of the equipment listed below may fall short of meeting your demand.

### AE Systems For Boats Consuming 6 Amp-Hours/Day

*Solar*—(Two) 10W panels, or one 20W, 30W, or 35W panels.

*Wind*—Small diameter wind units like the Aquair 50 or the LVM. Even the rigging-suspended large-diameter units can be set up occasionally to top up the battery, then stowed away again. Large-propeller pole mounts are not necessary for boats with a very light load.

*Water*—Any trailing-log type water generator will generate the required electricity if used occasionally. A free-wheeling transmission shaft on the engine can accept a *small* generator to provide charge when needed. No need to install an auxiliary prop generator for a very light load.

*Selecting, Installing and Operating Your System*

## AE Systems For Boats Consuming 14 Amp-Hours/Day

*Solar*—(Two) 20W or 35W panels, or one 40W or 44W panel.

*Wind*—Aquair 50, Ampair 100, LVM 50, Thermax or any of the rigging-suspended units that can charge the battery and be stowed out of way until needed again. Higher output large-propeller pole mounts are again not necessary with this size load, but can be rigged and used occasionally.

*Water*—Any of the water generator options. They need only operate when required to charge the battery. A voltage regulator is requisite for a separate through-hull prop that is always spinning.

## AE Systems For Boats Consuming 40 Amp-Hours/Day

*Solar*—(3) 35W, (3) 40W, (3) 44W, if sunshine and exposure is good; (4) 35W, (4) 40W, (4) 44W, if fair to good conditions exist.

*Wind*—Ampair 100*, Thermax*, and any of the large-diameter prop wind units, including Fourwinds, Friendly Energy, Hamilton Ferris, Redwing, Seebreeze, Silentpower, W.A.S.P., Windbugger, Winco (refer to average daily output chart for appropriate wind strengths for each).

*Water*—Aquair 50, when constantly sailing day/night with an average boatspeed of 5 knots (i.e., on long passages). For boats with an average speed of 4 knots or more, Aquair 100, Fourwinds, Friendly Energy, Hamilton Ferris, Powerlog, Redwing, Seebreeze, Silentpower, and W.A.S.P. are all adequate. Also an auxiliary through-hull prop and a generator on a free-wheeling prop shaft (e.g., Sailcharger) will work, provided the correct generator/prop combination is used.

## AE Systems on Boats Consuming 100 Amp-Hours/Day

*Solar*—Usually impractical by itself, but good in combination with high output wind or water systems.

*Wind*—Only four units are realistically able to regularly handle this kind of load on their own, according to manufacturers' ratings, and then only in moderate to windy climates with good wind exposure. They still might need an occasional backup to boost the battery. The Silentpower wind unit has by far the highest rated output at between 40–45 amps/hour in 25–30 knots of wind. The other three units are Fourwinds (any model) Windbugger and Friendly Energy. In good generating conditions Hamilton Ferris, Redwing, Seebreeze, W.A.S.P. and Winco could also satisfy a large load, although to ensure sufficient power they are better used in conjunction with another AE system.

*Water*—Most of the high-output water units on the market can meet a heavy load, provided the boat sails a steady 5 knots. This includes well-matched generators/alternators on auxiliary prop shafts and on free-wheeling transmission shafts. Refer to the average daily output chart for specific units.

---

* Provided they are used in windy climates with good exposure.

*Matching a System to Your Needs*

# CHAPTER 10

# OTHER ALTERNATE ENERGY SYSTEM COMPONENTS

Once you have selected an alternate energy system, there are a number of other AE components you should know about, components that distribute, meter, store and control the electricity you produce. This chapter lists those that are necessary, and those that are just nice to have, where to get them and how much they cost.

## BATTERIES

Batteries are necessary on any boat with an electrical load. They store the power you produce. The basic workings of a lead-acid battery are described in Chapter 2, including the difference between a surface charge, an automobile type that delivers a large current for short periods of time, and a deep-cycle battery that supplies small amounts of current for longer periods. To choose an appropriate battery for your AE system you need to understand battery construction and resulting characteristics.

The active material in a battery (see Chapter 2) is made of either pure lead, lead-calcium or lead antimony, each having different characteristics.

Lead-calcium and lead-antimony alloys are mechanically much stronger than pure lead grids. Most batteries on the market, both automobile and deep-cycle, have lead-antimony grids.

Batteries with lead-calcium grids have extremely low self-discharge rates (energy loss when not in use), as low as 1% of battery capacity a month. They also use little water and typically are sealed, or no maintenance units. Thus, these batteries are a favorite for use with photovoltaic systems, although they are not true marine batteries. These photovoltaic batteries are not tolerant of deep cycling and must be protected from a deep discharge by an automatic load disconnect. Often they only operate in the top 20% of the total battery capacity.

Deep-cycle lead-antimony grid batteries (marine, lift truck, golf cat, etc.) can handle deep discharge much more readily than the lead-calcium grid type, but have a high self-discharge rate, as high as 30% of the total battery capacity per month. High performance deep-cycle batteries, such as those used on fork lifts, generally are too heavy and costly to use on small boats. When a high-load golf cart or fork lift motor discharges a battery, the voltage drops to an unusable level before the battery is completely drained, thus protecting it. A small load like cabin lights will keep operating until the battery is completely exhausted, a situation that must be avoided if the battery is to last.

## Recommendations:

If you have a fairly constant source of electricity (from solar panels, a water generator on a passage, or a permanent or pole-mount wind unit), a light electrical load and have an automatic load disconnect to protect the battery from deep discharge, use one or two sealed, maintenance-free 100 amp-hour photovoltaic batteries with lead-calcium grids. Besides having a low self-discharge rate, they allow you to leave your boat for long periods without coming back to a dead battery. The Delco 2000 (available from photovoltaic dealers) is reasonably priced at about $135.

If one battery on board is used specifically for starting the engine, it should be the high cranking power type that delivers more power instantaneously.

If you prefer to deep cycle your battery rather than have an automatic load disconnect, and/or have a medium to heavy electrical load, your best choice is a deep-cycle lead-antimony grid battery. Many different models of this type of battery are available. Shop for the best price and warranty combination.

On *Kjersti* with a light electrical load and a solar panel as our primary AE system, we are in the photovoltaic battery category. We didn't know about them when we first installed our system, so we purchased a deep-cycle 100 amp-Sears DieHard marine battery that cost $80 and offered a 3-year warranty. It worked just fine.

Two batteries are on the market that are superior, if the manufacturer's claims are correct, in terms of deep-cycling ability, lifespan and warranty. These are the Everfair Enterprises EE 300, 8D and the EE 200, 4D.

## WIRING/TERMINALS

The electricity you generate should pass through properly sized and insulated wires. The battery terminals should be tight and protected from corrison. While some AE manufacturers include all the proper wiring, polarized connectors and terminals necessary for operation, most do not. The following list is a guide to what to buy if your unit does not come fully equipped.

### Wire Type

This should be annealed copper wiring with a stable, flexible, all-weather, UV insulation

|  | Amp-Hr. (20 Hr. Rate) | Cycle Life | Cost Retail | Size | Weight (lbs.) | Expected Battery Life (Years) | Warranty |
|---|---|---|---|---|---|---|---|
| EE 300, 8D | 225 | 2000 | $250 | 20½" × 11" × 9" | 160 | 12 | 10 Yrs. |
| EE 200, 4D | 175 | 2000 | $230 | 20½" × 8½" × 9" | 120 | 12 | 8 Yrs. |

*Other Alternate Energy System Components*

on the outside. The one exception might be a permanently mounted solar panel with wiring from the backside of the panel leading through the deck to the battery inside the boat. In this case, the stable, all weather UV coating isn't necessary.

## Connectors

A quick method of disconnecting the AE system from the battery is convenient when moving or stowing the unit. This connection should be polarized, so it can be assembled only one way. It's a necessity for DC systems, where plus and minus must be connected correctly. Use either a marine grade polarized connector or an exterior polarized connector, such as used on power tools and extension cords with a third wire ground. These are sold in electrical or hardware stores. Get the kind that comes apart and has wires wrapped around captive screws. A screwdriver is all you will need for assembling.

To make your system look good, use marine through-deck polarized fittings which allow you to plug in the unit without showing a long wire running to the battery. It's a nice

**Fig. 10-A.** A typical polarized connector.

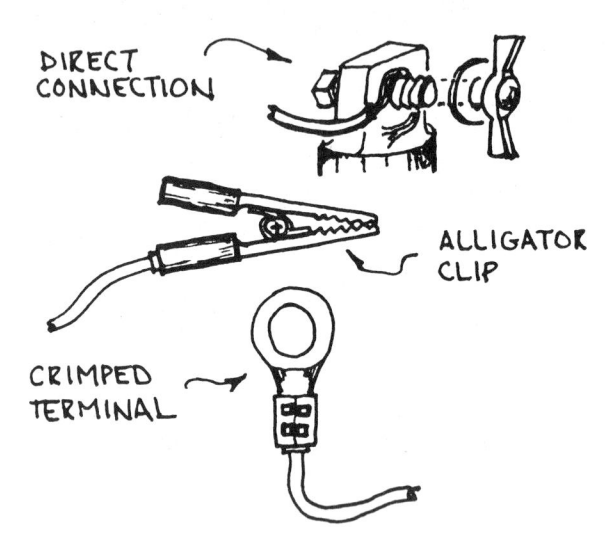

**Fig. 10-B.** Methods of attaching output leads to battery terminals.

esthetic touch. A.E.G. Telefunken offers nice through-deck fittings with its solar panels.

## Terminals

All AE systems are supplied with at least two external wires, so terminal connectors are not necessary at the generator end. The terminal at the battery can be twisted wire secured to the battery terminal or alligator clips that can easily be released. Regular crimped wire terminal lugs, correctly sized, that can be bolted to the battery terminal along with the other electrical leads, are the best option.

**APPROPRIATE WIRE GAUGE FOR A MAXIMUM OF 3% VOLTAGE DROP**

|  |  | \multicolumn{4}{c}{Length of Total Wiring Run (Feet)} |
|---|---|---|---|---|---|
|  |  | 20 | 30 | 40 | 50 |
| Maximum | 5 | 14 | 12 | 12 | 10 |
| Current | 10 | 12 | 10 | 8 | 8 |
| Flow | 15 | 10 | 8 | 6 | 6 |
| (Amps) | 20 | 8 | 6 | 6 | 5 |
|  | 25 | 8 | 6 | 5 | 4 |

*Selecting, Installing and Operating Your System*

Undersized wiring is not only a safety hazard but will also result in reduced generator output.

## DIODES

A diode placed in the positive lead to the battery allows current to flow into but not out of the battery. Without a blocking diode, current will flow out of the battery and into an AE device. Wind and water generators can become electric motors instead of generators and begin to *use* electricity from the battery. Solar panels act this way, but to a lesser degree. Make sure the diode, which itself is polarized, is oriented properly when it is installed.

While solar panels do *not* usually come with a diode, most manufacturers of wind and water units install diodes in their systems, often in the control box. Some wind units include a bump start switch (a questionable feature) that momentarily bypasses the diode and starts the unit in light winds. We've found if winds are that light, then very little usable energy is available, anyway. It's easy to purchase your own diode from an electronic shop, but keep two things in mind. Diodes are rated in maximum current flow in amps and by peak inverse voltage (PIV). This should be higher than the maximum battery voltage (or other charging source in the hybrid system). For example, for a double 35W panel solar system with a maximum current flow of 4.4 amps and maximum inverse voltage of 14V, you might select a 5 amp diode with a PIV of 25V.

Diodes also can be used to isolate your batteries while at the same time charging them equally.

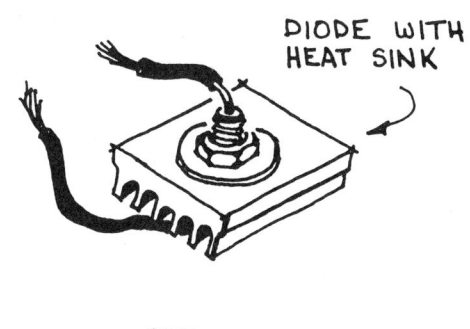

**Fig. 10–C.** Two common types of diodes used in alternate energy systems.

## METERS/INSTRUMENTS

While perhaps not necessary, meters are a convenient way to monitor your system. When we purchased our first wind unit, the manufacturer advised us to regulate the electrical load by generating whenever the lights began to dim. With two big batteries and sporadic use of electricity, the lights never dimmed, so we only raised the generator a few times during our first season on the boat. Consequently, it was a whole year before we discovered the unit wasn't working due to a defective generator. A meter would have demonstrated the defect immediately and saved us a lot of trouble. While this was an extreme case, we've met other AE system owners who discovered problems with their units by monitoring them with various meters. An added benefit of meters is they are

Other Alternate Energy System Components

fun to watch and make it easy to demonstrate the system to friends, relatives and other alternate energy enthusiasts.

## Ammeter

Two good places for ammeters are in the circuit from generator to battery to show your rate of charge, and in the wiring to your electrical panel, to show your rate of current drain. For accurate readings, match the top of the ammeter scale closely to the known maximum currents in your system. Always make the meter scale the higher of the two.

## Voltmeter

A voltmeter can be installed in the generator-to-battery circuit to display the charging voltage. As there is a small voltage drop through a voltmeter, they are usually not

**Fig. 10–D.** Our Hamilton Ferris control panel with voltmeter and ammeter.

**Fig. 10–E.** Spa Creek Instruments' battery condition indicator (courtesy of Spa Creek Instruments).

used in low-output photovoltaic and wind systems. A voltmeter can be useful, mounted at the electrical panel to indicate state of battery charge.

## Battery Condition Indicator

This is much easier to read than a voltmeter and gives the state of charge directly, rather than converting it from voltage or hydrometer readings. The meter shown in Fig. 10-E can be used independently or it can replace your existing electrical panel voltmeter.

## Anemometer

A windspeed indicator is convenient for comparing wind velocity to wind generator output, especially since comparing these two elements is how the manufacturer develops an output curve or each system. When deciding whether to buy a hand-held or masthead anemometer, remember that a wind generator permanently mounted at the masthead may experience quite different wind condi-

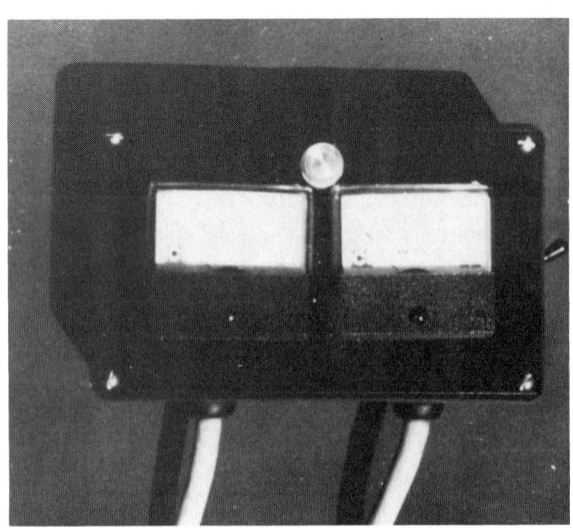

*Selecting, Installing and Operating Your System*

tions from a unit 10 feet above the water. But hand-held anemometer are easy and convenient to use. A very accurate, well-made anemometer, the Sims BT, is carried in the Coast Navigator catalog for $72.50. Airguide manufacturers a hand-held windspeed and direction indicator for about $50. A simpler, but less accurate hand-held model that gives rough estimates is offered by Dwyer Instruments for about $8–$10.

## Knotmeter

Like the windspeed indicator, the knotmeter provides a means of comparing boatspeed to generator output to determine the output curve of a water unit. Many boats already have a knotmeter. If not, you can calculate your speed from time and log readings.

## Voltage Regulator

A voltage regulator is not necessary if you have a huge battery capacity compared to your generator output. It's also not necessary to have a regulator if you fastidiously monitor your battery state of charge to prevent overcharging and if you disconnect your generator when you're away from the boat for long periods.

While not a necessity, a voltage regulator is convenient, particularly if you have a high output generator that might overcharge the battery. A charter captain we interviewed aboard a boat with a Fourwinds D72-3 wind generator assumed his load was at least as high as his generator output. He didn't have a meter, and so did not regularly check the battery condition. As a result, two very expensive batteries were overcharged and ruined. A voltage regulator would have prevented the overcharging. Regulators also allow you to safely leave your boat and still keep the battery at full capacity. Obviously, this arrangement only applies to self-tending AE units.

Most alternate energy manufacturers offer a voltage regulator for their equipment. Some solar companies combine it with a battery condition indicator, circuit breaker and load disconnect. Voltage regulators are sized by the maximum current expected to flow through them, or the equivalent to your maximum generator output. The Thermax Corp., One Mill St., Burlington, VT 05401, offers a good regulator for small output generators (up to 10 amps) for about $70. Prices go up to as high as $200. Check with your generator supplier for recommendations.

**Fig. 10–F.** Solar Power Corp. voltage regulator, with integral battery condition indicator and circuit breaker (courtesy of SPC).

## AUTOMATIC LOAD DISCONNECT

This device automatically disconnects any non-essential load before the battery reaches deep discharge. In combination with a voltage regulator, it is a convenient method for automatically protecting and maintaining your battery. Arco's Solar Battery Protector, rated at 20 amps, provides charge control, automatic load disconnect with warning light, and a place to interconnect wiring between the panels and photovoltaic batteries.

## EMERGENCY SHUTDOWN SWITCH (WIND UNITS ONLY)

This is a simple toggle switch, usually mounted on the control box, that functions as an on/off switch. It is used for quickly stopping the unit in an emergency or for keeping the propeller of a rigging-suspended unit from rotating until the generator is raised into position. It should not be used for routine stopping, as the generator will experience wear on the brushes and/or commutator. This is not an adequate stopping device on its own in strong winds.

## FUSE OR CIRCUIT BREAKER

A fuse, rated slightly higher than the maximum output current, should be included in the output leads from a generator or solar panel to protect the battery in the event of a short circuit. On wind units that aren't self-tending, it will also protect the generator from producing potentially high current levels in strong winds. But, if the fuse is blown by high winds, or if the circuit becomes disconnected for any other reason (e.g., the polarized connector comes apart) the load on the generator is cut, meaning its propeller speed can increase by as much as 50%. This is a potential runaway situation, and one very good reason why an easy, practical method of manually or automatically stopping the propeller is a necessity.

# CHAPTER 11

# OWNER-BUILT ALTERNATE ENERGY SYSTEMS

Many sailors wish to build their own alternate energy system for their boat. If you are considering building from scratch, make sure your motives are well-defined and your skills are equal to the task. An astounding number of owner-designed and built AE systems end up costing more and generating less electricity than an equivalent manufactured unit. On the other hand, many sailors have built successful systems. If done properly, building your own can be a rewarding experience.

Detailed instructions on how to build an AE system are beyond the scope of this book. What is appropriate here is to advise on what is best done by the amateur and what work should be left to the professionals. The real issue is, what's the most practical way to save money.

## BUILDING YOUR OWN SOLAR PANEL

The best way to buy solar cells is through a national distributor/wholesaler, two of which are:

Energy Sciences Inc.
16728 Oakmont Ave.
Gaithersburg, MD 20877

Solec International Inc.
12533 Chadron Avenue
Hawthorne, CA 90250

Using Energy Sciences' 1984 retail prices for solar cells to evaluate the cost of an owner-built 30W panel we have:

| Materials | Cost |
|---|---|
| thirty-six, 4″ diameter solar cells | $270.00 |
| mounting board | 5.00 |
| wiring | .50 |
| solder and flux | 1.00 |
| silicon | 6.00 |
| transparent cover | 5.00 |
| frame | 4.00 |
| exterior connectors | 1.00 |
| | $292.50 |

Strictly on economic grounds, it's hard to compete with some of the standard solar panels, considering product quality, labor and lack of warranty. The cost may be low compared to some of the more expensive marine panels, but home-built systems can't be walked on and don't have the professional appearance of manufactured units. In commercial solar panels, temperature, humidity and contaminate levels (dust, etc.) are carefully controlled during assembly. The units are factory-sealed to prevent environmental damage. They are backed by a long warranty period, something unavailable when purchasing individual cells. Energy Sciences also offers a complete kit for assembling a 20W panel for $199.

It's fun and challenging to build your own solar panel, provided saving money is not your only consideration. If it is, you're better off buying a manufactured panel from a reputable solar panel dealer. Used and factory seconds also are available, but often difficult to locate. Energy Sciences often sells Solarex seconds on a first-come, first-serve basis.

If you are interested in building your own panel, good construction details are available in the January 1982 issue of Cruising World magazine. If you decide to buy a ready-made panel, you still can put your skills and creativity to work to save money.

## Installation

Mounting of the panel and wiring, including the diode and meters, usually is left to the owner. Here is a list of things you can easily do to customize your panel and make it work better. See Chapter 12 for more details.

- Build a mounting system that allows the panel to swivel, rotate, tilt-up or be moved to better face the sun. One sailor mounted his 35W panel on a pole at the stern on a universal swivel. This works very well, but should be used only at anchor. Another sailor made an effective gimbaled frame (similar to a galley stove gimbal) mounted off the stern that allowed the panel to swivel 90° either way from horizontal (see Fig. 12–E).
- Design and build creative mounting systems so the panels won't diminish your boat's appearance.
- Install the diode and ammeter/battery condition indicator flush with a bulkhead (interior or exterior) so it blends in with other instruments.
- Conceal the wiring by installing marine-

**Fig. 11–A.** Example of light reflectors for a solar panel. They can be hinged so that they fold out of the way when not in use.

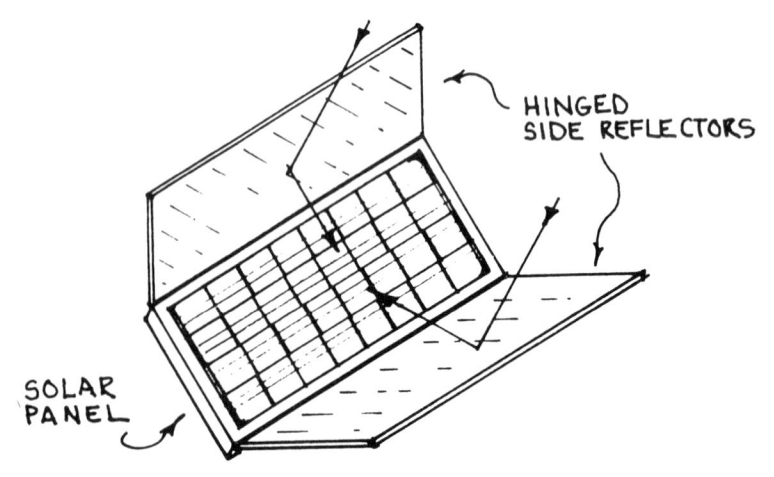

grade polarized deck connectors underneath or near the solar panel. Then continue the wiring below decks to the battery.

## Reflectors

A simple way to increase panel output by 20% or more is to make reflectors that direct more sunlight onto the panel surface. In effect, you are increasing the intensity of the sunlight that reaches the cells. This technique also is used in solar heating applications. But it becomes more important in photovoltaics because the cost per square foot is much higher. Make lightweight reflectors from polished aluminum (best) or even smooth, thin boards painted with a highly reflective white paint. They should be hinged to the panel along one edge and tilted to get maximum sun reflection.

## Wood Border

It's easy to dress up a standard solar panel with a thin border of teak or other nice wood. Screw or epoxy the border to the frame, or make it part of the mounting system on your deck or cabin top (see Chapter 12).

## MAKING YOUR OWN WIND GENERATOR

More sailors experiment with wind generators than any other AE system. Making a wind unit, like building a solar panel, can be a fulfilling project. But unless you are very skilled and are knowledgeable about aerodynamics and generator/alternator design, you will do best to purchase a generator and the propeller to go with it from a wind unit manufacturer or supplier. Manufacturers use special types of generators or alternators to achieve high output at a lower RPM. They carefully select a prop that works most efficiently with their generator. Few of the single-piece wood props are made by wind generator manufacturers. Usually they are produced by a company specializing in propellers, indicating the complexity of accurately making a high-performance prop that complements a particular generator.

Last winter we met three sailors who had purchased Hamilton Ferris generators and made their own propeller blades in an effort to save money. The results were widely varied. The first sailor built an aluminum prop/hub assembly. Theoretically, his idea was fine. But he lacked the proper tools and as a result turned out a heavy and ill-performing propeller. Finally he bought a Hamilton Ferris prop. The second sailor fashioned a lovely prop that looked as if it would perform well. It turned out to be substantially heavier and noisier. We tried both blades on the same generator and hooked up our ammeter. The professionally-made propeller had twice the output. The third sailor produced a blade with very good performance. It started in slightly lighter winds than our own unit and seemed to have an equivalent output. It was, however, considerably noisier. His success was due to his skilled background and perseverance. He carved model airplane blades for a hobby and this particular prop was his fifth attempt.

*Owner-Built Alternate Energy Systems*

If you try to make a propeller, make sure the whole prop and hub assembly is properly balanced. Place the assembly on a shaft and slowly turn it, checking for improper weight distribution. Blade material can be removed or weights added to achieve the correct balance. It also should be balanced (same as your car tires) to make sure it will rotate smoothly at high speeds. One sailor's hand-carved prop developed a wobble during its trial run in windy conditions and self-destructed before it could be stopped. As an additional safety measure, make your propeller highly visible. One owner, who made his prop out of clear acrylic, later walked into it while the unit was operating and was injured.

Don't get discouraged. But do give the project the consideration it deserves. While wind units are fairly simple machines to build, they require specific knowledge, skills and tools to work satisfactorily. If, by making your own propeller and generator combination, you spend half the money otherwise spent on a manufactured pair, and the resulting electrical output is only half as great, then you might or might not be satisfied. But if you want the highest output possible from a system, your money is better spent on a professionally manufactured and matched propeller and generator.

Generator combinations may be purchased from Hamilton Y. Ferris II Co., Redwind, and John Betus, P.O. Box 4661, Princeton, FL 33032. See Chapter 17 for other addresses and additional information.

If you purchase a professionally made propeller/generator combination you still need to mount the unit. There are many things you can do on your own during installation.

## Rigging Suspended Mounts

*Frame*—A frame is needed to hold the generator/propeller assembly in place and to suspend it high in the foretriangle. Wood, 813 stainless steel or aluminum are the best materials. Make sure you have a secure method of attaching the generator to the frame. We know of one homemade unit that was damaged when the generator crept backwards out of the supporting cradle until the propeller hit the frame. In another case the generator/prop assembly wobbled so badly the owners were afraid to use it (and rightly so). One good way to secure the assembly is to support it with a cradle locked to the generator with one or two screws. Make the frame just longer than the blade to simplify hanging and to protect the prop when the unit is lowered. The hanging generator and prop assembly must be well balanced or it will tip resulting in a hazard and a loss of efficiency.

**Fig. 11–B.** A welded cradle to secure the generator for pole mounting.

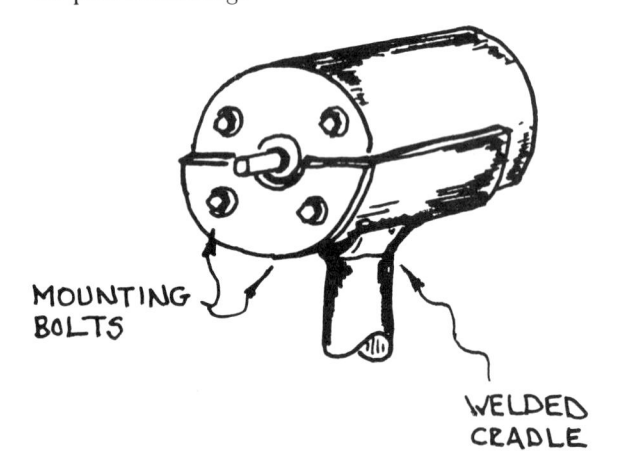

**Fig. 11-C.** A well constructed, owner built frame for a rigging suspended wind generator.

*Swivels*—Use 813 stainless steel or bronze swivels at either end of the frame to allow the unit to rotate smoothly.

*Tail Fin*—The tail fin needs a large enough surface area and an arm long enough to allow the unit to rapidly track the wind. The shape is less critical. We've seen triangles, rectangles and arrow-like tails. A high aspect ratio shape is one of the best both in performance and appearance.

*Turning Handle*—This supplies an easy way to manually rotate the unit out of the wind. It should be a sturdy arm attached to the middle or lower part of the frame and angled down and back towards the rear of the unit, within easy reach.

*Tie-Off Line*—With a small line attached to the end of the tail vane or turning handle, you can tie off the unit at 90° to the wind or keep it from rotating more than 30° or so when in operation. Use 1/8" nylon cord.

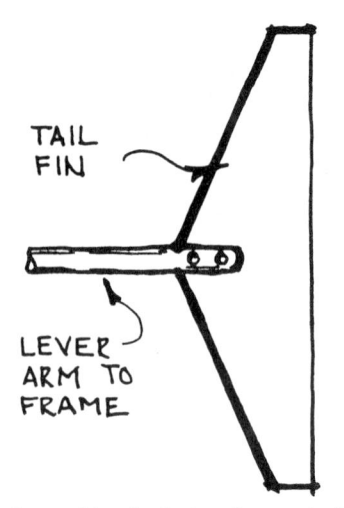

**Fig. 11-D.** A good basic design for a wind unit tail fin.

## Pole Mounts

A favorite with home AE builders is to mount a purchased generator and propeller on a pole and frame. However, many of these units do not look very good. With a little more effort they can have the finished polish of a professionally-made unit.

The frame holds the generator/prop assembly securely in place and provides a base that will accept the proper size pole so the generator and propeller will be balanced when mounted on the pole. This means a 1½″ diameter base for a 3½′ diameter propeller or a 2″ diameter base for propeller diameters 4′ or larger. A well-lubricated bearing arrangement allows the unit to effectively track the wind. The frame provides a firm point to attach the tail vane.

The frame may rotate on the pole or the whole top section of the pole may rotate with the frame. Slip rings or brushes to transfer the output current are optional if you want to eliminate external wiring.

Rig tail fins and turning handles in the same fashion as discussed in the previous section on rigging-suspended units. Pole installation is the same as for manufactured units (see Chapter 12).

## Fixed Mounts

The simplicity of a fixed mount makes it another favorite with home AE builders. A wind unit that is rigidly attached in position will be less efficient than one that can swivel (see Chapter 18). Large-diameter propellers on fixed mounts must have a reliable means

**Fig. 11-E.** An owner built, pole mounted wind generator aboard the yacht *Destiny*.

**Fig. 11-F.** A wind generator with overspeed braking device fix-mounted to the mizzen mast.

of being stopped in strong winds. A effective and automatic method of limiting propeller RPM (like the Wincharger air brake) is difficult to build, so most sailors rely on manual brakes. There are three fixed-mount locations for a generator/prop combination:

*Mast Mount*—A cradle type of frame that securely holds the generator is attached to the forward part of the mizzen mast (or main mast on a catboat). The generator is bolted in place and a manual brake cable led down to an accessible location.

*Bow Pulpit Mount*—Hanging above the bow like a modern figurehead, these mounts are rigidly attached to the pulpit rail. They require a miniature wood or metal bow sprit that extends the unit forward and high enough to avoid any interference with anchor lines. The output wires are led back to the battery, preferably below decks. See Fig. 3-P in Chapter 3.

*Shroud Mount*—This is possible on boats with double inner shrouds, and preferably is located on the mizzen mast. A generator support, resembling a support for belaying pins, is strapped at a convenient location across the shrouds above head height. The generator is attached to this support. This arrangement cannot be installed on main mast shrouds and operated while underway if there is any chance of the jib or jib sheets touching the prop.

## MAKING YOUR OWN WATER GENERATOR

### Trailing Log Type

The best kind of water generator for the home AE builder is the type with the generator mounted on a gimbal at the stern (instead of towed in the water). The generator can be purchased from a wind/water manufacturer and the propeller bought new or used from your local outboard dealer. Other parts include:

*Gimbal Mount*—It doesn't have to be fancy, but it should hold the generator securely, be fastened tightly to the deck, and allow the trailing line a fair lead regardless of the motion or speed of the boat. Painted aluminum or stainless steel are the best materials.

*Propeller Shaft*—A 3′ long stainless steel shaft, usually 9/16″ diameter—depending on propeller used—should be attached to the prop. Make a wood, plastic or metal cone to slide over the shaft and rest against the prop hub to lessen the resistance when it moves through the water. Install a shackle on the forward end of the shaft for attaching the towing line. A bowline will work but a splice has a more finished appearance.

*Towing Line*—Use approximately 75′ of 7/16″ braided nylon line or equivalent. If your boat is slow, try 40–50′. If it is fast, you may need to use as much as 100′, unless a device like Hamilton Ferris's diving plane is employed.

*Generator Shaft Attachment*—An easy way to attach the forward end of the towing line is to make a stainless steel hub with a ring welded to the outboard end. Secure this to the shaft with a couple of set screws. As the torque generated is relatively high, provide indented seating for the screws in the shaft to prevent losing the propeller assembly. On long passages, use Loc-Tite compound or equivalent to prevent screws from working loose. Then tie or splice the tow line to the welded ring.

## Auxiliary Generator on Freewheeling Shaft

*Auxiliary Propeller Shaft through the Hull*—Both of these systems are good for home AE builders. See basic installation procedure in Chapter 12.

# CHAPTER 12

# INSTALLING THE SYSTEM

Every manufacturer includes instructions on how to assemble and install their equipment. Because alternate energy companies are usually small, it's easy to get to know them and, if needed, obtain special mounting advice. This chapter will explain what is involved in the installation of an AE system and familiarize you with the necessary parts, connections, tools, etc., for installing today's production AE equipment. Included are suggestions that are not included in installation sheets from the manufacturers. To get the rated performance from any AE equipment, it must be properly mounted. With the appropriate tools and skills, most sailors are able to install the system themselves.

## SOLAR PANELS

As a sailor, you know the sun's highest point in the sky each day occurs at solar noon on a north/south axis. Since a solar panel's maximum output occurs when it is perpendicular to the sun, the installation of a photovoltaic system for a house is straight forward. The panels are faced due south and pitched (depending on latitude) to an angle off the horizon that best intercepts the sunlight. On a boat, the process is less certain. The south-facing part of the boat depends on your heading. It's also difficult to pitch a panel off the deck where it won't snag a sheet or line. If you pitch it too far, it might face the sun perfectly on one tack but be totally blocked on the other. You need an installation that will give you a good average output throughout the day, regardless of the point of sail or position at anchor. Several ways exist to achieve the desired output.

If possible, leave the solar panel(s) free to be moved around the sunny areas of the boat. By moving our panel about the boat to achieve near maximum output throughout the day, we've found that one 35W panel satisfies our entire electrical load (with the exception of running lights) when cruising in sunny climates. It *does* make a difference if you keep the panel in direct sunlight. A permanently-mounted panel will not give the same high output. Remember that most cruising other than an ocean passage, is spent at anchor or day-hopping in easy winds. Both of these conditions require moving panels around a boat a lot to obtain maximum output.

It's always a good idea to provide a method of temporarily securing your movable panel. This is absolutely necessary for offshore sailing, and depending on the motion of your boat, a practical means of keeping the panel in place and out of the way while sailing in exposed waters. The best way to get good average outputs with this arrangement is to mount it flat on your deck or cabin roof, or clamped to your stern rail. A quick method of securing and removing solar panels is a definite advantage. Most sailors we've met without a quick disconnect system wished they had incorporated such a feature.

## Deck Mount

Most panels come with a thin (1"–1½") aluminum or wood frame around the perimeter. A few, such as the Solec Solarcharger, Arco M82, and Free Energy Systems' panels, provide corner mounting holes in place of the frame. Regardless of type, several rules must be followed when mounting solar panels.

Don't attempt to bend a solar panel to match the curve of your deck. Despite some manufacturers' claims, this can stress the individual solar cells and eventually cause cracking. Instead, make wooden runners that support the panel above the curve, mounting it so the panel's smallest dimension spans the curve. Most solar panels are 12–14" wide and 3–4' long, making them easy to support above a curved deck. PCD Labs and A.E.G. Telefunken have mounting systems that permit installing as many as four panels side by side resulting in a graceful curve much like the sections of a bow window.

You might have to mount your panel on a double curve if the deck slopes in both directions. Unless the curve is very pronounced, this should present no problem. Just taper the runners in one direction. Solar panels come in various shapes, so check to see which ones best fit your boat. While the larger panels offer a better price per watt, it sometimes is easier to fit two smaller panels than one large one.

Air must be allowed to circulate freely under panels. Don't seal the panel to the deck around its entire perimeter. If air circulation is inhibited, the panel temperature will rise and the electrical performance will drop. When installing a narrow panel on runners along the long sides, leave the ends unsup-

**Fig. 12–A.** Multiple panel solar installations (courtesy of PDC Labs and A. E. G. Telefunken).

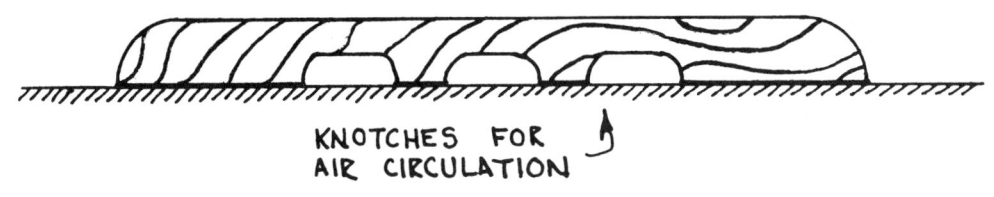

**Fig. 12-B.** Notched wood runners for deck mounted solar panel.

ported so the air can move freely underneath. With a panel that's supported all around, notch the runners to provide breathing spaces. Kyocera panels include small feet at each corner of the trim, a handsome feature that also provides adequate ventilation.

## Quick Disconnect

As discussed earlier, the best panel mounting arrangement is one that provides some means for quick and easy removal. Notch the runners, as shown in Fig. 12-B, or make them out of two pieces, providing a shelf for the panel to rest on. Pivoting wood fasteners adds a nice wood trim around the perimeter and allows the panel to be removed in seconds. For Kyocera or A.E.G. Telefunken panels, where a mounting system is built into the wood trim, you only need to add four small blocks (one each side) with a pivoting fastener on each. This locks the panel in place and provides for easy removal.

A quick-removal mounting arrangement probably won't have the clean appearance of a permanently-mounted panel with wooden plugs to hide the screw or bolt holes, and might be more susceptible to theft. You must decide which is the most practical mounting method. Adding an electrical connector

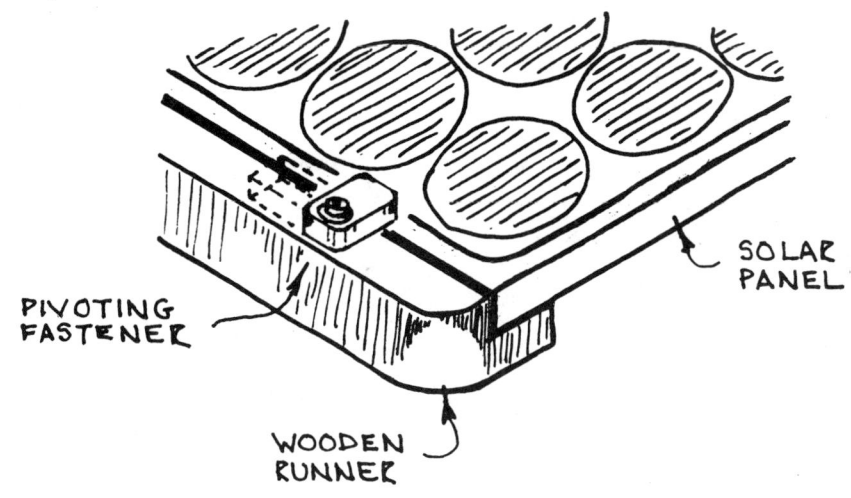

**Fig. 12-C.** Turnscrews used for securing a movable solar panel.

*Installing the System*

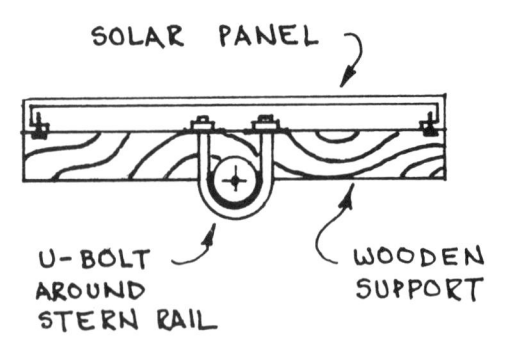

**Fig. 12-D.** Method for mounting a solar panel on the stern rail.

under the solar panel minimizes the clutter. Conceal the wiring (all panels have connections coming out the back) unless you need to move the panel around the boat.

## Stern mount

*Stern Rail*—A sturdy stern rail mounting gives your panels plenty of air circulation, flexible exposure, and protection from the tread of feet and loose sheets. First make two rotating supports out of wood, cut to width of the solar panels. Then clamp them to the rail with U-bolts. See Fig. 12-D. To facilitate rotation, separate the two mounts along a straight section of rail, keeping them a few inches or more from either end of the panel to spread the weight and increase the strength. Put a bit of rubber or tape around the rail to protect it from stratching. Then tighten the U-bolts just enough so you can tilt the panel back and forth. A panel mounted either side of a self-steering vane or transom-hung rudder is a popular configuration and very attractive. PCD Labs and Arco both offer stern rail mounts.

*Stern Gimbal*—A small gimbal arrangement held off the stern with wooden brackets (like small davits) allow one or more solar panels

**Fig. 12-E.** Gimbal mounted solar panel on the stern of a catamaran.

to swivel 180° to catch the sun. Panels on this mount are out of the way and generally get maximum exposure. This arrangement is illustrated in Fig. 12-E, and is a good alternative for boats with no stern rail or one taken up with other cruising gear.

## General Notes

Keep the length of wire from panel to battery as short as possible to minimize energy losses. Use the wire gauge recommended in Chapter 10.

Most standard solar panels come with exposed terminals or a junction box, for ease of wiring multiple panel arrays. Seal them with a silicon adhesive or equivalent, making sure it can easily be removed if new connections need to be made (don't use hard-setting epoxy).

Diodes are mounted in the positive wire to the battery. Be sure to mount the diode facing electrically in the proper direction. If you are unsure, test it with a test lamp or voltmeter. Don't take polarity for granted. We used our first solar panel for a week before we realized it had come from the factory with the wrong markings on the diode wires. An ammeter to measure output would have helped us spot the problem immediately.

Multiple solar panels should only be connected in parallel in a 12V system (i.e., join positive wires together and negative wires together). The wires can be joined at the panel or lead separate wire sets to the battery.

A voltmeter is not used in a solar panel output circuit because it causes a voltage drop and some current loss. An ammeter is preferable. It uses no current, and should be wired in series with one of the output leads. Make sure to purchase a meter with the correct scale for your output current.

If a voltage regulator is not used, have a reliable method of regularly checking the battery state of charge.

## WIND GENERATORS

### Pole Mounts

When buying a wind unit to mount on a pole, you will need to complete the installation yourself. All wind generator manufacturers offering a pole-mounted version include a generator/propeller frame that accepts a pipe. But the pipe, stays, struts and deck fittings usually are supplied and installed by the customer. An optional pole mount kit is available with the Ampair 100.

Most companies include a yaw thrust bearing in their frame so the unit easily swivels to hunt the wind. Some offer slip rings and brushes to transfer the output current at the yaw axis without twisting the wires (see Chapter 17 for product descriptions).

For these units, a fixed pipe of the correct diameter and wall thickness can be inserted into the mounting and secured to the boat with the wiring inside the pipe. Fourwinds does not supply either, suggesting instead that the pipe to which you attach the frame rotate inside the PVC pipe strapped to a stern rail stanchion. We've seen several of these units, and while they certainly are functional, they do not rotate smoothly and often don't look good—especially those with external wiring twisted around the pipe. Some manufactured units do not include a

*Installing the System*

frame for pole mounting. To be well supported, a large propeller unit should have a 1½–2″ diameter T6061-T6 or equivalent schedule-40 aluminum pipe. We feel the Fourwinds, at 1¼″ diameter, is too small. For the smaller units, Ampair, LVM and Thermax, 1½″ diameter schedule-40 aluminum pipe is recommended. While galvanized steel pipe can be used, it is heavy, could adversely affect your compass, rusts, and is not nearly as attractive. Polished aluminum and stainless steel poles similar to those used for Biminis and cover structures on large boats give the best appearance.

Once you have the pole, a way for the unit to rotate with the wind, and a neat wiring arrangement, you still need to attach the pole to the boat. Use a threaded deck fitting to accept the end of the pipe. Add a rubber pad between the fitting and the deck to cut down on noise and vibration, and some struts or stays for structural support. See Fig. 12-F.

*Deck Fitting*—Use either a standard galvanized pipe flange with an inner diameter large enough to accept the outer diameter of the pole, or construct one of wood with a hole drilled out for the pole and a means of holding the bottom of the pipe securely in place.

*Rubber Pads*—Oddly enough, these are rarely used to cut down on the annoying noise and vibration transmitted below decks which is the major complaint of pole mount owners. Install them at the bottom of the pole and at all stay or strut connections. They should be sufficiently tough to withstand wear, but have enough resilience to cut down on vibration.

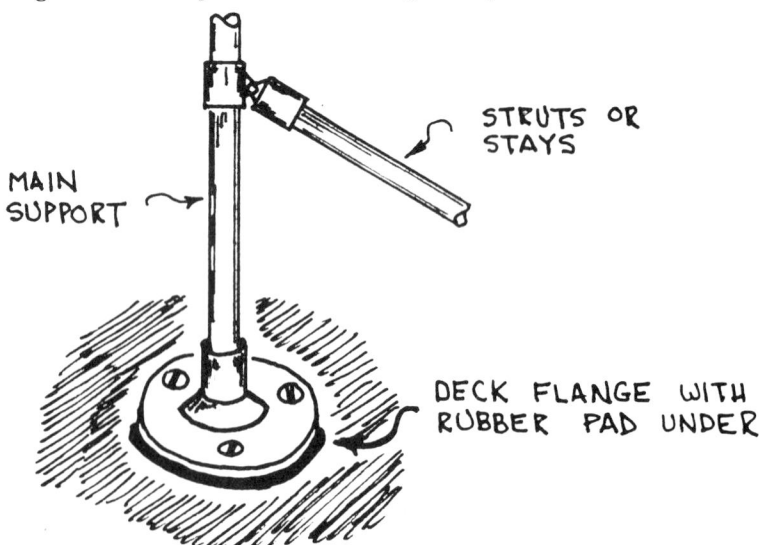

**Fig. 12-F.** Example of a deck fitting for a pole mounted wind unit.

*Selecting, Installing and Operating Your System*

*Stays or Struts*—Wire stays with small turnbuckles to adjust and level the unit can be used to steady the pole. Three are required to keep the pole rigid. Only two aluminum struts are needed. For a very clean appearance, try the rod and sockets that are used for heavy duty Bimini construction. In either case, the stays or struts should be attached to the pole no higher than half the propeller diameter away from the center of the generator shaft. This means that on a pole with a 5′ diameter prop, the supports will start minimum of 2½′ below the generator shaft, leaving the upper part of the pole unsupported. This is the reason we favor strong poles when using large diameter wind generators.

## Masthead Mounts for Small Wind Units

These types of mounts are similar to pole mounts in that they are installed on short poles. The masthead rig is only possible if there is no other equipment aloft to interfere with the rotating wind generator. As an example, the turning radius of the Ampair 100 is 18″, so an area 36″ in diameter must remain clear for complete rotation of the unit. Masthead running lights are likely to be partially blocked by the supporting pole, and antennas may also get in the way.

A radar mount extends the unit on a strut at the upper forward part of the mizzenmast. The strut must be strong and well secured to the mast and must extend just beyond the turning radius of the wind unit. Small-diameter propeller units all have slip rings and brushes to transfer the electrical output, so there is no need for exposed wiring. Run the wires through the short pipe and down the inside of the mast to the storage batteries.

## Rigging Suspended Units

With these machines there is nothing to install permanently. What goes up comes back down after operation. You do need an exposed forward stay (or topping lift for mounting in the main triangle) for securing the upper part of the unit. You will need a separate stay if roller furling is rigged. You will need a halyard for raising the unit and swivel fittings at either end of the frame for a 360° rotation. Two, preferably three, lines are needed at the bottom of the frame, tied off port, starboard and forward to keep the unit in place. Finally, it's necessary to have someplace on deck to tie the lines. If nothing is available, install deck cleats or rings at the necessary locations.

The first time you set up the unit, experiment to get the correct halyard setting and tie-down system. The wiring from the generator must be kept clear of the rotating blade. We run our wiring through a loop in the tie off line at the bottom of the frame, keeping it tight and out of the way. These units are designed to be frequently set up and taken down, so it's best to keep the lead wires (with control and diode, if possible) permanently attached to the battery. Adding a polarized quick disconnect fitting allows you to disconnect and stow the wind unit without touching the battery terminals. Plug in the connector when you're ready to generate. Our wind/water unit came with this convenient arrangement. We unplug the wind unit and plug in the water unit when we want to switch systems. The actual operation of a rigging-suspended unit, including setting up and taking down, is covered in Chapter 13.

### General Notes

When mounting a control box with an ammeter and voltmeter—either supplied by the manufacturer or self-made—choose an accessible location so you can easily monitor your output.

When installing the wind unit, make sure it has good propeller clearance and can rotate 360°, including space for the tail vane.

Always mount the unit so the rotating blades are well above head height to avoid any chance of injury.

If a voltage regulator is not used, make certain you are able to regularly monitor your battery's state of charge. This is especially important for large prop wind generators.

## WATER GENERATORS

### Trailing Log Type

The only part of this unit that needs to be permanently installed is the gimbaled frame at the stern. With the exception of the Powerlog unit, which trails its generator in the water, this frame is supplied by the manufacturer. The installation is simple. Drill several holes through the deck or other structure, fasten down the gimbal with the correct size bolts or screws, and seal the holes. If you like, a polarized electrical deck connector can send the wiring from the frame to the battery without going through or around the cockpit.

When selecting a location for the gimbal, remember that the trailing line must always have a fair lead. The entire line spins rapidly and requires some swinging room off the stern. We know of a line that was neatly severed on a self-steering gear when the boat had to do some fast maneuvering. The propeller walks slightly to port, so try to mount it on the port side, if possible, away from any taffrail log or fishing line.

### Auxiliary Prop Through the Hull

Once you have selected your generator/alternator and propeller combination, install the propeller shaft through the hull as you would any engine drive shaft. Unless you are very familiar with this work, professional help is required. Be sure the shaft is accessible for periodic maintenance or emergency repair. The exact size of your shaft will depend on the propeller you select, but will probably be 9/16". Bearings, mounted on metal struts on either side of the hull to hold the shaft in place while rotating, must be very accurately aligned to allow the shaft to spin freely. A small standard stuffing box will seal the hull where the shaft passes through. It should take much less abuse than an engine shaft stuffing box due to the relatively low RPM with slight vibration. The generator shaft can be directly connected to the propeller shaft by a universal joint or similar arrangement, or by means of a belt and pulleys. See Fig. 3-U.

A belt and pulley is excellent because the generator shaft is better isolated from the prop shaft. It's also much easier to disengage the generator using a simple manual clutch to loosen the pulley belt. The drive system resistance depends on belt type, tightness and pulley ratio, which in turn depends on the RPM necessary for your particular generator. The best generator is one that gives a high output at a low RPM and a pulley ratio of 1 or 2. If you find an inexpensive generator/alternator that requires a higher ratio, you might find it works well enough to justify the money saved. Only experience with your own generator and boat will determine the right combination.

## Auxiliary Generator on Freewheeling Prop Shaft

Install a bracket near the transmission shaft that will hold a generator securely. Mount the generator so its shaft is parallel with the transmission shaft. Place a pulley on each shaft directly opposite each other. Their ratio will depend on your specific generator. Again, only by testing your generator on your boat will you determine the correct shaft speed. It is important to have both an electrical and a mechanical way to disengage the auxiliary generator/alternator from the transmission shaft when the engine is in use. Everfair Enterprises makes a custom mounting system for use with any marine engine shaft. An electrical disconnect protects the battery and generator at high RPMs. A mechanical disconnect prevents the generator from spinning uselessly and incurring wear.

## General Notes

Because of the relatively high electrical output of a water generator whenever the boat is underway, a voltage regulator or a very accessible means of regularly monitoring the battery state of charge is recommended.

*Installing the System*

# CHAPTER 13

# OPERATING AND MAINTAINING ALTERNATE ENERGY SYSTEMS

A few precautions are necessary to protect both you and your AE unit. Special attention must be lavished on large-prop wind generators, since they are the most dangerous of all AE systems. Preventive maintenance is necessary to keep any generator operating at peak performance.

## SOLAR SYSTEMS

Since solar panels are sealed and have no moving parts to maintain, little is involved in their operation. In fact, you can't even tell if they're operating without looking at a meter. Still, there are a few things to do and remember:

Regularly wipe clean the surface of the panel with a soft damp cloth to remove dirt, salt deposits or bird droppings. This will keep your panel operating at maximum output. We use leftover cotton diapers.

If your panel has a glass cover, be careful not to hit it with sharp objects. Impact strength is high for tempered glass, but it will fracture easily once a crack is started. Our SPC panel passed a Jet Propulsion Labs impact test that simulated the repeated impact of a 1.5″ diameter hailstone traveling at 100 MPH.

Check to make sure nothing in the rigging is blocking the panel—such as flags, laundry, pennants, etc. A friend of ours noticed one day that the needle on his ammeter kept jumping around. Upon further investigation he found his fluttering ensign, attached to a backstay, periodically covering part of the panel.

Be careful when maneuvering around the solar panel during wet conditions. Wet glass can be very slippery, even if the glass is textured.

Kyocera offers a protective shield that can be placed over its solar panels. You can make one yourself if your panel is mounted where it is vulnerable to damage.

Regularly check the electrical connections for corrosion. Corroded or loose connections will reduce or stop output current.

With most manufactured panels it is impossible to disassemble the unit to troubleshoot electrical failures. But the rate of failure is extremely low. If you experience an internal electrical malfunction it is best to return the panel to the factory.

If you leave your boat unattended for long periods of time, solar panels have sufficiently high voltage (16–18V) to overcharge your battery. The exception is the Arco M63 that only has 30 cells and therefore cannot produce enough voltage to harm the battery. On cloudy days, however, the lower voltage may be a hindrance. Charge up your battery before leaving the boat and then either disconnect the panels during your absence or install a voltage regulator.

## WIND SYSTEMS

### Self Tending Units

By definition, any of the self tending wind generators, whether suspended on poles or in the rigging, do not require attention during operation. Fixed units are stopped by turning the boat out of the wind or by using a manual brake. For general operation and maintenance, see the section, General Notes On All Wind Generators, appearing later in this chapter.

### Non-Self Tending Units

SETTING UP RIGGING SUSPENDED UNITS

1. Attach generator and tail vane to frame if not already attached while stowed.
2. Secure hub and prop to generator shaft. Set screws should be tight, but take care not to strip screw heads.
3. Attach the upper suspension line and lower tie-down lines to the swivels on the frame. The length of these lines is determined during initial installation.
4. Secure the upper part of the suspension line to the shackle at the end of a jib halyard, and then shackle this connection to the front stay to keep the pulling force in line with the stay. An alternative arrangement is known in Fig. 13-A. We can't do this on *Kjersti* because our prop is longer than the frame and might hit the line to the stay.
5. Next, secure generator output wires to

**Fig. 13-A.** Methods of attaching a rigging-suspended wind generator to the front stay.

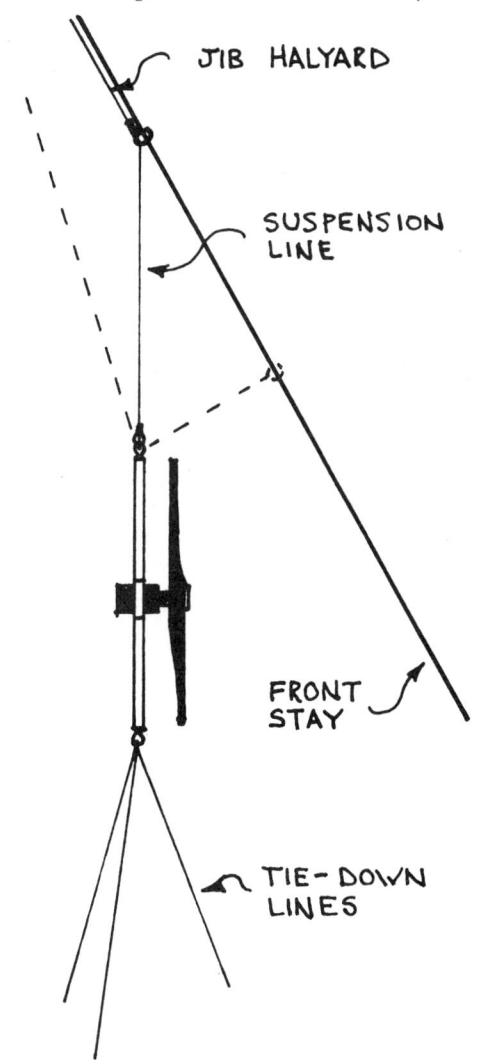

*Operating and Maintaining Alternate Energy Systems*

the frame to keep them clear of the rotating blade. We usually just pass the wire through a loop in the tiedown lines.

6. The prop must not rotate until the unit is hauled up and in place. Someone can hold onto the prop until it is in place, or tie off the prop until the unit is raised. A more sophisticated device is an electrical shutdown switch that will stop the prop in an emergency. Turn this switch off while raising the unit, then on again when you are ready to generate. the Silentpower unit has an optional manual disc brake that can be pulled when raising the unit, which will shut down the unit no matter how hard the wind is blowing. Bicycle-type manual brakes won't work on large prop wind units when the wind is above 25–30 knots.

7. Check to make sure the unit is sitting level fore and aft and side to side. Adjust tie-down lines as necessary. A rigging-suspended mount may have a tendency to jerk back and forth in chop. Some sailors use lead weights suspended from the tail vane to dampen the motion.

## Lowering Procedure

1. Stop the prop completely before lowering the unit. Turn the unit 90° out of the wind by its turning handle or a small line attached to the tail vane. A bicycle-type manual brake is sufficient if the windspeed is not over 25–30 knots.
2. Once the prop is stopped, ease the halyard until the unit is resting on the deck, making sure the prop is not the first part to touch down.
3. Disassemble and stow in reverse order to the setting-up procedure.

All non-self tending wind units without overspeed governors should have a reliable method of turning them out of the wind. This is your only dependable and safe means of getting the blade to stop. Pole and rigging-suspended mounts should have a permanent

**Fig. 13–B.** Silentpower wind generator suspended in the foretriangle.

handle or tail vane tie-off line. On the Fourwinds, the top section of the pole rotates as part of the frame and owners we've met say they turn it out of the wind by hand or, if the wind gets too strong, with a pipe wrench. But emergencies happen too fast. Don't depend on a nearby tool. If leverage is needed (and in strong winds it most certainly will be), it should come from a strong, well-attached handle on the frame or a line attached to the end of a well-supported tail vane.

We have seen homemade units fix-mounted on a bow pulpit, on the mizzenmast and in the mizzen shrouds. While these areas are fairly out of the way, these units cannot be turned out of the wind without turning the entire boat, and should therefore have a foolproof manual brake. We've seen a number of other techniques employed, such as squeezing the hub with a bare hand, pressing a piece of wood against the hub, and laying a hand protected in a welder's glove against the spinning prop. These methods are very dangerous, especially in strong winds.

A line attached to a handle or tail vane can serve an additional useful function. As the wind increases, the unit (pole or rigging mount) can be rotated incrementally out of the wind to keep the prop rotating at a safe speed. We usually start with 30°, then 45°, then a full 90° to stop the generator completely if conditions warrant it. With the Fourwinds, you can adjust the pitch of the prop blades. One setting is for light winds, the other for heavy.

The emergency switch on the control box (if included) should be reserved for emergencies, or for raising a rigging suspended unit. It should not be used for routine stopping, which will cause premature wear of the brushes and bearings.

A non-self-tending wind unit *should not* be left in operation when you are away from the boat. Good judgment is necessary. Obviously, in fair weather, you can go ashore for short periods of time and leave your unit unattended, especially if you remain close enough to row out should the wind pick up. But if there is even a slight chance of a squall or strong winds, save yourself the worry and tie off the unit.

Don't, for any reason, work on a rotating wind unit while it's in operation.

## GENERAL NOTES ON ALL WIND GENERATORS

Always make sure the prop and hub assembly are properly secured to the generator shaft. The fasteners have a tendency to back out with use, so make periodic checks. One day our blade started making an unusual sound. Upon inspection we found our set screws had either backed off or we hadn't tightened them properly in the first place.

Since the generator shafts are stainless steel and *most* hubs are anodized or painted aluminum (Ampair uses a stainless steel insert), the resulting slight corrosion can cause the prop to stick on the shaft. A touch of Teflon grease on the generator shaft before mounting the prop will prevent the hub from sticking and allow it to be removed easily. Also, grease around the front bearing to help keep moisture away.

Most of the airplane-type propellers are made of Sitka spruce for its strength and light

**Fig. 13-C.** Pole mounted wind generator with an unmatched propeller.

weight. The leading edge will wear a fair amount (from bugs, rain, etc.) unless it is adequately protected with epoxy paint, abrasion-resistant tape or some equivalent material. Get recommendations from your propeller manufacturer. The leading edge must be kept smooth so the prop will start up in light winds. We know one sailor who epoxied a thin layer of aluminum foil on his blade's leading edge. The resulting surface was just rough enough to keep the unit from performing, as it had before, in light winds.

If you ever have to replace your prop, stay with the same type (i.e., same diameter,

126
*Selecting, Installing and Operating Your System*

pitch, airfoil). They usually are carefully matched with a generator to produce a rated output. Sailors we met on a boat in Eleuthera in the Bahamas had broken their Fourwinds blade and replaced it with an inexpensive, ultralight airplane prop. Their new output was exactly half what it had been with the original prop.

Many manufacturers sell a factory balanced prop and hub assembly, which should not be disassembled. Balancing a high-speed prop is important and some manufacturers sell a tool so you can do it yourself.

Internal bearings support the generator shaft on each end. Occasionally a bearing will wear out, but it's a simple job to replace them. Your manufacturer can supply assembly details on your particular unit. Spare bearings can be purchased from the manufacturer or from a supplier. The first time your generator is disassembled, have it done at an armature shop so you can learn how it's done. The bearings might stick to the shaft and require a gear puller to remove them.

If you have a permanent-magnet generator, your brushes (either two or four) will require replacement about every 2-3 years. This is also a good time to replace bearings, as the unit is disassembled. Field current alternators have slip rings and rotating contacts for the field excitation current and

**Fig. 13-D.** A disassembled permanent magnet generator.

require less frequent replacement (about every 5 years). Permanent magnet alternators do not have rotating field current contacts to maintain.

If the generator output leads become disconnected from the battery during operation, either by coming unplugged or blowing a fuse, there will be no load on the generator and the prop RPM can increase by as much as 50%. Since no current is flowing, the generator will not be harmed, but in strong winds the prop and frame of a non-self-tending rig could be overstressed.

## WATER SYSTEMS

### Trailing Log Type

STARTING PROCEDURE

1. Mount the generator on its gimball at the stern.
2. Tie the forward end of the towing line to the generator shaft and *neatly* coil the remainder of the line, including prop and shaft.
3. Connect the generator wiring to battery terminals (or plug into the polarized connector).
4. Make sure there are no obstructions on the boat or in the water and that the trailing line has a fair lead (watch your hands and feet).
5. Lower the prop into the water and rapidly pay out the coiled line. With practice, this can be done in one quick motion.

During operation, the unit is so quiet you wouldn't know it was working without looking at the spinning line. On long sails, remember to check your battery state of charge regularly. As battery voltage increases, the amperage output of the generator decreases, but still is capable of dangerous overcharge. Once your battery is fully charged, pull in the unit and stow it until needed again. This will save wear on your generator and eliminate any unnecessary drag.

### Retrieving the Unit

Two effective techniques will stop the spinning prop so the unit can be hauled aboard.

**Fig. 13-E.** Our conversion kit that turns a wind generator into a water trolling generator.

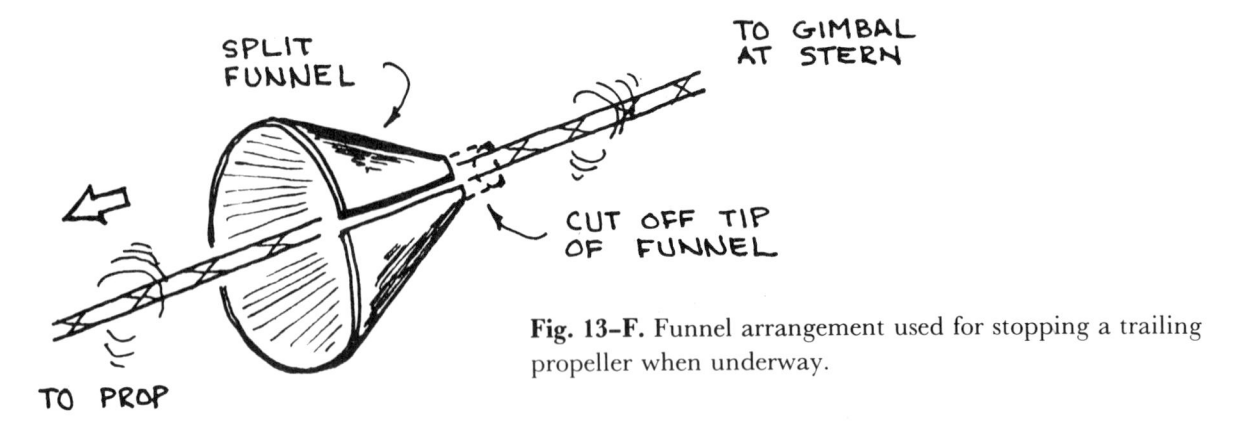

**Fig. 13-F.** Funnel arrangement used for stopping a trailing propeller when underway.

Stop the boat by coming into the wind. Pull in the prop and shaft, then get underway again. Another trick, which we learned from Hamilton Ferris and which is now endorsed by many water genertor manufacturers, is to take a large plastic funnel and split it from end to end. Place the funnel over the spinning line (large end toward the water). Let the funnel fall down the line. It will hit the spinning prop, stop the water flow and thus stop the prop from rotating. The prop and shaft can then be hauled in as you continue to sail. The funnel must be heavy duty to stand up to this treatment. Perhaps one day water generator manufacturers will offer this as an optional piece of equipment.

*Note:* Operating the Powerlog water generator is similar to other trailing-log types except that the prop *and* generator are put in the water. The tow cable does not rotate, only the prop, which is directly connected to the generator in the water. Make sure the inboard end of the cable is secured to a strong cleat or equivalent before tossing the assembly in the water. The electrical leads are contained in the tow cable.

### Auxiliary Through-Hull Propeller Shaft

Because of the quiet, hidden nature of this generator and its potentially high output, it is wise to invest in a voltage regulator. That way you never have to worry about overcharging and it will electrically disconnect the unit when the battery is fully charged. This type of system is totally self-tending.

Regularly check the stuffing box for leaks and wear. For best performance and least drag, keep the prop free of barnacles and other marine growth. Occasionally check the shaft bearings for wear.

If a belt system is used, check the belt for wear and correct tension. Keep a spare belt on board.

### Auxiliary Generator on a Free-Wheeling Propeller Shaft

Electrically and/or mechanically disconnect the generator from the battery before the engine is turned on or if motoring for a long time (more than entering or leaving a harbor). Check belt tension and wear and keep a spare on board.

*Selecting, Installing and Operating Your System*

A voltage regulator is advisable unless you check your battery state of charge often.

## General Notes on Water Generators

Most generator considerations are the same as for wind units, except that more frequent bearing and brush replacement may be necessary if the generator is run continuously during long passages. And, as always, batteries should be fully charged before leaving the boat unattended for long periods of time.

*Operating and Maintaining Alternate Energy Systems*

# CHAPTER 14

# TAKING YOUR SYSTEM (AND PHILOSOPHY) ASHORE

A surprising number of cruisers who made their escape on a sailboat become thoroughly disillusioned with their live-aboard existence. Often, the reason for this disillusionment is the drastic change in lifestyle. How can anyone expect to be happy using 5 gallons of fresh water a day when on shore their consumption was 100 gallons? Can you be as happy using 1200 watt-hours of electricity a day as you were in a home that consumed 20,000 watt-hours a day?

If you can maintain a simple, economical lifestyle on land that includes your essentials for a good life, then you will have an easy time adapting to *any* mode of living, whether it's cruising on a sailboat, camping, backpacking or just living in a home on shore. Potential live-aboards who first experiment with a simpler lifestyle on shore learn at less expense how well suited they are for living aboard their boats.

The key to success lies in learning how to think small, how to live with less, how to be more self-sufficient and self-reliant. If you can live that way in a home, you can live that way on a boat. The opposite also is true.

Given a successful experience living on a boat, a move ashore is no reason to disregard everything you learned afloat. Whether you are a live-aboard or weekend cruiser, your cruising philosophy can be taken ashore and applied with equal success.

The actual AE system you installed on your boat can be set up to supply electricity to a home ashore. Solar panels and wind generators are easily adaptable for shore-based use. Even the generator from a water system can be turned by a wind propeller or by a geared water turbine powered by a stream. If you require more electricity than a boat AE unit can supply, simply upgrade the system by installing more solar panels or a larger wind generator.

Alternate energy technology has come of age. Existing photovoltaics and other solar technology is such that every new home or building in the U.S. should:

- Provide a south-facing roof area at the correct pitch for the present or future installation of photovoltaic solar panels. Trying to retrofit a system to an existing house can be difficult and expensive.
- Include a solar heating system, mounted on that south-facing roof, to supply domestic hot water. (California already has laws to this effect.) Other forms of active solar systems should be employed where practical.
- Make maximum use of insulation. When building codes require more insulation,

**Fig. 14–A.** A typical solar-powered navigational aid one of the most useful, yet unused marine applications (courtesy of A. E. G. Telefunken).

**Fig. 14–B.** Self-sufficiency at its best—bringing your system ashore.

designers will be encouraged to add inexpensive passive solar techniques—as a minimum—in all new construction.

Take advantage of windpower in areas with good average windspeeds. As illustrated on boats, these systems do not have to be large and costly if your energy requirements are reduced.

Also, let's encourage the use of hydroelectric power where it is feasible. Some small projects have minimal environmental impact and are surprisingly effective.

Support government involvement in alternate energy technologies, especially through tax incentives and funding for continued development. But even more important than government funding is adopting a national attitude that this is indeed the direction our energy policy should be heading.

The same attitude and opportunities are equally applicable to the boating industry. Boat designers and builders should include one or more alternate energy options on their boats. Alternate energy devices should be encouraged for all remote marine energy use sites such as lighted buoys, lighthouses, etc.

Pass on to friends, relatives and acquaintances what you have learned about alternate energy. Help other people learn from your experiences. That's the way small movements become large ones. We sailors have a unique opportunity to experiment with alternate energy, to demonstrate to others its effectiveness and to further its cause on both land and sea.

# PART IV
# Reference Section

# CHAPTER 15
# GLOSSARY OF ELECTRICAL TERMS

**Alternating Current**—The type of electric current found at your dock or in your home. In this case, the current travels first in one direction and then in the other. Each complete set of reversals is called a cycle. The number of cycles a second is called the frequency of the alternating current. The commonest frequency in the United States is 60-cycles. The modern term for cycles-per-second is Hertz.

**Alternator**—A type of generator that produces alternating current; this is then rectified, by diodes in the unit, into direct current that may be used to charge a storage battery. Current is created by rotating a magnetic field on a shaft (rotor) inside some wire windings (armature) or a stationary housing (stator).

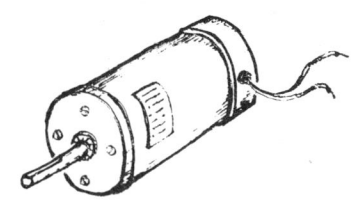

**Ammeter**—An instrument for measuring the current in an electrical circuit. Many different scales are available depending on circuit current flow to be measured.

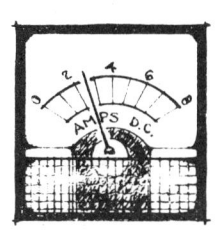

**Ampere**—A measure of the current in an electrical circuit. One ampere is equivalent to the flow of $6.24 \times 10^{18}$ electrons per second.

**Ampere Hour**—The measure of total current flowing in one hour. If a lamp drawing one amp was left on for one hour, it would consume one amp-hour of electricity. This term is primarily used to rate batteries. A typical marine battery would have 100 amp-hour of capacity (often referred to as ampacity).

**Appliance**—Anything on board a boat powered by electricity.

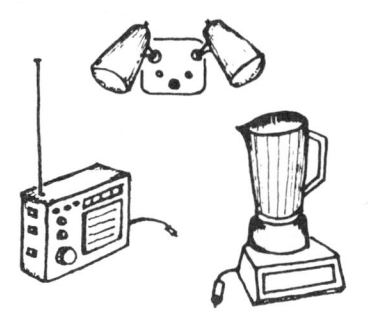

**Battery**—A group of two or more cells that produces electrical current (DC) due to a chemical action. The lead-acid type, rated at 12V (6 cells in series at 2V each), is widely used on boats and in automobiles.

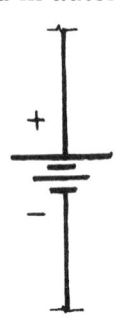

**Circuit**—A simple circuit is comprised of a source of voltage (battery), a resistance to current flow (an appliance or load), and a conductor (copper wire) connecting the two. Usually included in a boat circuit are a switch for on/off control, and a fuse or circuit breaker as a safety device in case of overload.

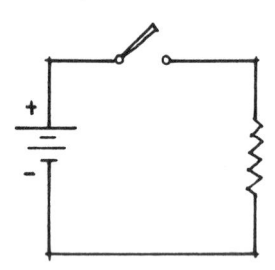

**Closed Circuit**—Often a source of confusion to students of electricity, it refers to a circuit that is continuous, or complete, and allows electrons to flow freely. A closed switch is ON, an open switch is OFF!

**Circuit Breaker**—A safety device, often an automatic switch, that opens or disconnects a circuit if more than a set amount of current flows. Unlike a fuse, which must be replaced, a circuit breaker may be reset once the trouble is corrected.

**Conductor**—Any material that contains many available drifting electrons. When under the influence of voltage from a battery, the direction of electron drift is from the negative pole to the positive pole. Metals are very good conductors, and copper wire is the usual conductor in an electrical circuit because of its superior performance and low cost.

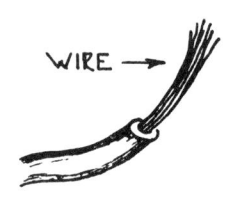

**Continuity**—A complete, or closed, circuit where electrons flow freely. Continuity is lost if a switch is opened, wires become loose, or an appliance is faulty (e.g., a light burns out).

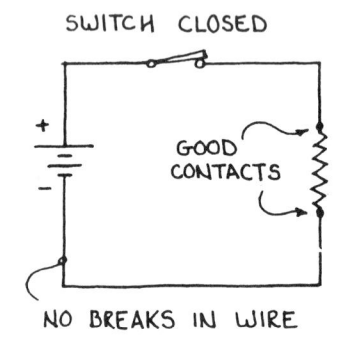

136
*Reference Section*

**Current**—The rate of drift or flow of electrons through a conductor. It is measured in amperes (amps) and is analogous to the flow of water inside a pipe. Current is governed by the voltage pressure and its resistance in the circuit (see Ampere).

**Diode (Blocking Diode)**—A diode is a small solid-state device that allows current to flow only in one direction. A diode is rated by the maximum current it will safely accept and by peak inverse voltage (PIV), the highest voltage that would cause current flow against the diode block. On boats a diode acts both as a conductor, allowing generated electricity to flow into the battery, and as an insulator, preventing the flow of current into the generating equipment, where it is simply lost or where it can damage equipment.

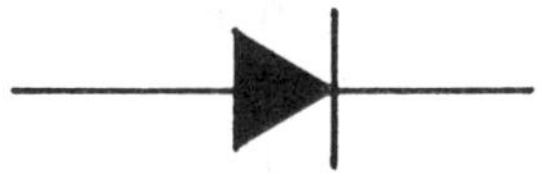

**Electricity**—The net drift or flow of electrons through a conductor. When a voltage induces electron drift, an energy wave is produced moving through the conductor near the speed of light.

**Electrolyte**—A material whose atoms or molecules become ionized in solution (see Ion). In a lead-acid battery, the electrolyte is a sulfuric acid and water solution. When ionized, sulfuric acid gives rise to one sulphate ion carrying two negative charges, and two positively charged hydrogen ions.

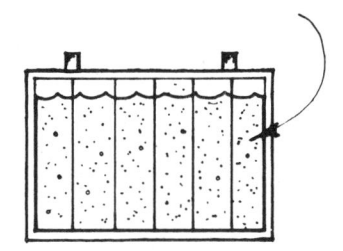

**Electron**—A basic particle that orbits rapidly around the nucleus of an atom. The total positive charge of the nucleus determines how many electrons are in orbit. Each electron carries one negative charge. Electrons revolve in various orbits; those in the outer orbit, or shell, can drift from atom to atom. These electrons cause conductivity as they move under the influence of voltage. The movement of electrons is known as electricity. Conductors have three electrons or fewer in the outermost shell; insulators have five or more.

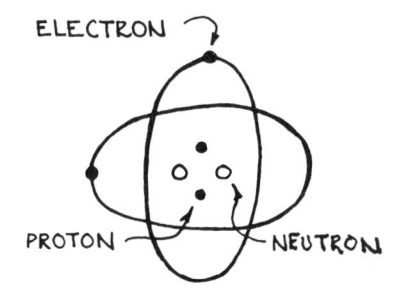

**Energy**—The amount of power produced by a generator or consumed by an appliance in a given amount of time (typically an hour). It is measured in watt-hours.

**Frequency**—the number of times an event recurs in a given period. In alternating current it is the number of times that the electrons reverse direction in one second.

*A Glossary of Electrical Terms*

**Fuse**—A safety device that protects an electrical circuit or appliance from too much current draw (overload). When the current through the fuse exceeds the rating of the fuse (ratings are in amperes or decimal amperes) the metal wire inside the fuse melts, opening the circuit before more damage occurs.

**Generator**—A rotating electro-mechanical device that produces electricity by rotating wire windings (armature) on a shaft (rotor) inside a magnetic field that is set into a stationary housing (stator). The motion of the wire inside the magnetic field initiates a force (voltage) that pushes electrons down the wire, creating electrical current. Although a solar cell is also a generator (DC only) it is *not* included in this definition.

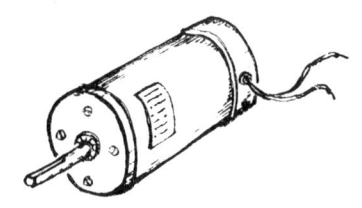

**Hydrometer**—A device used to measure the specific gravity of a liquid, such as battery electrolyte. Some units measure specific gravity with floating balls. Others give direct numerical readings on a float.

**Insulator**—Any material that does not allow electricity to flow freely, such as wood, rubber, plastics and glass. Insulators have five or more electrons in the outer orbit of each atom. They isolate the electrical charge in wires and electrical components, preventing them from leaking electricity, damaging each other or injuring anyone coming in contact with them.

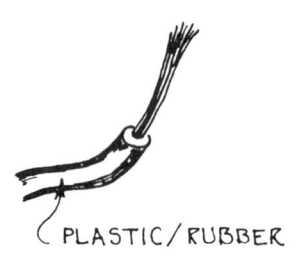

PLASTIC/RUBBER

**Ion**—An ion is an electrically charged atom or molecule formed when a neutral atom loses or gains one or more electrons. The loss of an electron results in a positively charged ion called a cation. The gaining of an electron results in a negatively charged ion called an anion.

**Load**—The power (in watts) consumed by an electrical device. It is related to the resistance of the device. The more devices or appliances connected in parallel, the lower the total resistance and the greater the load.

**Magnetic Field**—The force that surrounds magnets, and all current-carrying conductors. If there is relative motion between a conductor (e.g., copper wire) and a magnetic field, an electrical pressure, or voltage, is produced in the wire, causing electrons to flow. This is the principle on which the generator is based.

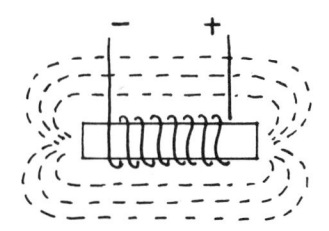

**Multimeter**—An instrument that functions as an ammeter (usually *low* amperage), voltmeter, ohmmeter and continuity tester.

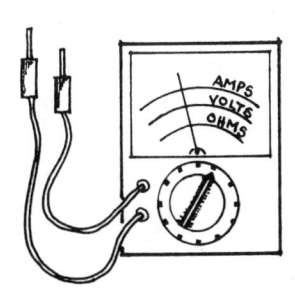

**Negative**—One kind of electrical charge, refers to an atom that has gained one or more electrons to become a negatively charged ion, and to the terminal on the battery from which electrons flow in an electrical circuit.

**Ohm**—The measure of resistance (R) in an electrical circuit. It is related to the voltage (V) and current (I) by the relationship V = I x R. Appliances have an inherent resistance and some components are rated in ohms. Corroded wiring, contacts or loose connections can cause excess resistance in a circuit.

**Ohmmeter**—An instrument for measuring the resistance of an electrical circuit or device. An internal battery supplies a small voltage that creates a current proportional to the resistance in the item being measured. The resistance readings may be taken on one of several scales.

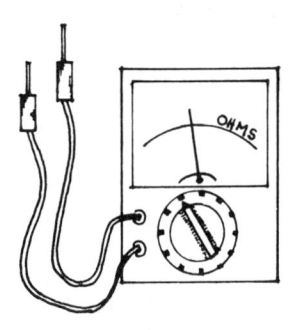

**Open Circuit**—An electrical circuit that does not allow current to flow because of an open switch, loose wires, or an appliance that has been removed from a plug or socket (e.g., a light bulb).

**Overload**—An excessive amount of current draw in an electrical circuit.

**Parallel Circuit**—Connecting electrical devices "head to head" and "tail to tail". If two 12V batteries are connected in parallel (positive wires together, negative wires together) the voltage will remain the same but the ca-

139
*A Glossary of Electrical Terms*

pacity (amp-hours) will be the sum of the two. The boat's instrument panel represents electrical appliances wired in parallel. Each appliance has its own switch and safety fuse or circuit breaker.

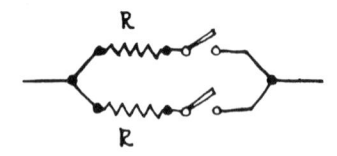

**Permanent Magnet**—A magnet that does not require electrical circuitry to maintain its magnetism. Such magnets are installed, for example, inside some generators.

**Photovoltaic Conversion**—The process that converts light (radiant energy) directly into electrical energy by means of a solar cell.

**Polarity**—The property of an object containing opposite forces. A magnet has north and south magnetic polarity. A battery that supplies direct current has a polarity due to its positive and negative terminals. All DC circuits have positive and negative leads, and are thus polarized.

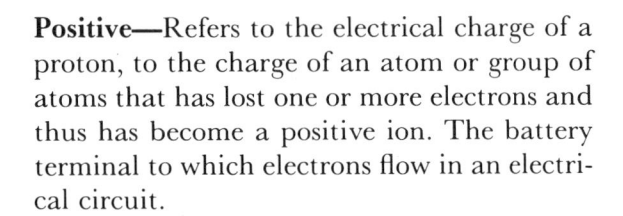

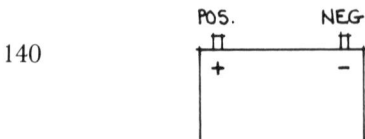

**Positive**—Refers to the electrical charge of a proton, to the charge of an atom or group of atoms that has lost one or more electrons and thus has become a positive ion. The battery terminal to which electrons flow in an electrical circuit.

**Power**—The rate at which work is being done by electricity. It is a function of voltage (V) and current (I); $P = V \times I$, and is expressed in watts. Generators are rated in watts according to their power output, and appliances are rated in watts according to the power they consume.

**Rectifier**—An electrical device, usually consisting of a configuration of diodes, that converts alternating current (AC) into direct current (DC). Alternators have a built-in rectifier since they must supply direct current to be stored in the battery.

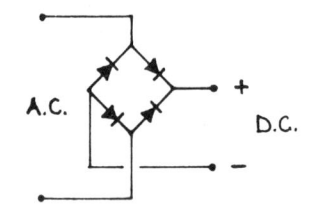

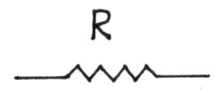

**Resistance**—The opposition to the drift, or flow, of electrons along a wire. It is measured in ohms, and is analogous to the opposition to the water flow due to pipe diameter and friction. The larger the diameter, the less the opposition. Large pipes or wires offer low resistance. All electrical components and appliances place a resistance on an electrical circuit.

**Revolutions Per Minute (RPM)**—The number of times in one minute that an object revolves 360°. Used to describe the rate of rotation of an engine, generator or propeller shaft. It is measured with a tachometer.

**Series**—Refers to electrical devices that are installed "head to tail" in an electrical circuit. If two 6V batteries are connected in series, positive terminal of one to negative terminal of the other, the voltage will increase to 12V, but the capacity in amp-hours will remain the same as one battery. If a switch is opened in a series circuit, current will stop flowing to all devices in the circuit.

**Solar Cell**—A thin wafer, usually a slice of a silicon crystal, that has been deliberately treated with impurities to create an excessive negative or positive charge. The size of the cell determines the amperage (typically .063 amps for a 4" cell). Solar cell voltage, regardless of size, is about one half volt.

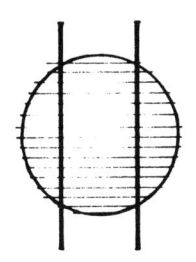

**Solar Panel**—A thin, rigid frame that houses solar cells connected in series to yield the rated power in watts. The size of the solar cells determines the amperage, and the number of the cells determines the voltage, typically 16–18V in a battery-charging application. For example, a 35W panel may have 36 cells of 4" diameter producing approximately 2.2 amps.

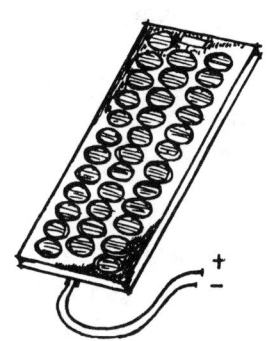

**Specific Gravity**—The ratio of the weight or mass of a given volume of a substance to that of an equal volume of another substance. For liquids, water is used as a reference with a specific gravity of 1.0. The specific gravity of the electrolyte in a fully charged lead/acid battery is between 1.26 and 1.3.

**Switch**—An electrical device that opens or closes electrical circuits. Many varieties and ratings are available.

**Volt**—The measure of the electrical pressure (or voltage) in an electrical circuit.

**Voltage**—The force that moves electrons along a wire in one direction. It is measured in volts, and is analogous to the pressure that

141
*A Glossary of Electrical Terms*

moves water through a pipe. Voltage on boats comes from one or more lead-acid batteries that usually are rated at 12V.

**Voltage Regulator**—An electrical control device that adjusts the output voltage from a generator before it reaches the battery. On car and marine engines, it drops the charging amperage from the alternator soon after start-up so that the batteries will not be overcharged. A special type of voltage regulator may be used with an alternate energy generator (although not necessary) so that batteries will always be protected from excessive charge.

**Voltmeter**—An instrument used for measuring the voltage in an electrical circuit or battery.

**Water Generator**—A device that converts the motion of a boat through water into electrical energy. A water propeller (similar to a propeller on a small outboard motor) rotates a tightly wound line as it is pulled through the water. The rotary motion of the prop and line is transferred to the shaft of the generator.

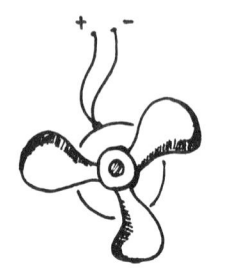

**Watts**—The measure of power being generated or consumed in an electrical circuit. Watts (P) is related to the voltage (V) and current (I) by the relationship P = V x I.

**Watt-Hour**—The measure of electrical energy generated or consumed in one hour: 1000 watt-hours = 1 kilowatt-hour.

**Watt-Hour Meter**—An instrument that measures the total energy produced or consumed in an electrical circuit. It is the electric meter outside your house or at a dock that indicates the number of kilowatt-hours of energy you consume. Not a practical device for on board use.

**Wind Generator**—A device that converts the mechanical energy of the wind into electrical energy. A propeller transforms the horizontal movement of an airstream into rotary motion. The greater the propeller/generator shaft speed, the higher the electrical output.

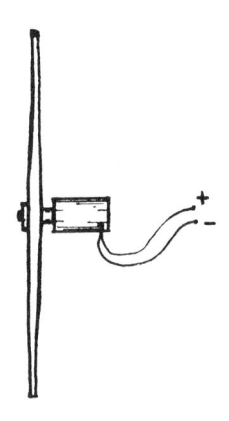

# CHAPTER 16

# A LITTLE THEORY

In Chapter 2 we presented the basic laws governing electricity in a circuit, placing particular emphasis on the formula for power that determines the rate at which we use and produce electricity. In this section let's discuss the *theoretical power* we can extract from the sun, wind and water, and incorporate into this theory actual conditions encountered on a boat. Let's see exactly how the devices that produce this power for us actually work.

## PHOTOVOLTAICS

Unlike the theory behind using wind and water to generate electricity, where the difficult task is to discover how much power is available for our use, we already know that the available power from sunlight is approximately 1 kw per square meter at the Earth's surface on a clear day. Not all this solar power can be converted into usable electrical power. Solar cells now available can convert 10% or more, or are said to be 10% efficient. At this efficiency, a square piece of land 100 miles on a side, experiencing average sunshine and covered with solar cells could produce the same amount of electrical power as the present generating capacity of the entire United States! Viewed another way, for each proposed 1,000-megawatt nuclear power plant we could substitute solar cells covering a square piece of land 5 miles on a side, the equivalent to the combined area of the south-facing roofs of 900,000 homes. The photovoltaic cell converts the sun's light energy directly to electricity. Since the relationship between light and electricity is not obvious, we need a little more background.

## Light

Light is visible radiant electromagnetic energy, most of which comes from the sun. Any light, however, will trigger the photovoltaic effect. This is why a solar calculator, for instance, will operate at night under a desk lamp. Other forms of radiant electromagnetic energy are radio waves (received by your AM/FM/SW/VHF), infra-red radiation (given off by any heat source), ultra-violet radiation (what attacks your sails and skin), x-rays and gamma rays.

Heinrich Hertz first discovered what we know as the photovoltaic effect in 1887, but not until Einstein put forth his Quantum Theory of light was the phenomenon explained. According to Einstein's theory, light consists of "quanta" of energy, now known as photons. The energy (E) of each photon is equal to h x f, where f is the frequency of the light and h is Planck's constant ($6.6 \times 10^{-27}$ erg-sec.), which states that radiation frequency is related to its quanta of energy. Visible light comes to us in a variety of colors, each having a specific range of frequencies.

Although E=hf is a very small amount of energy, it is present in each photon. When you consider that a trillion-billion photons are emitted each second from a 100-watt light bulb, you begin to appreciate how much power is available from bright sunlight.

**Solar Cells**

The solar cell itself consists of silicon, made from sand (the most abundant substance on Earth, after water). First, polysilicon is produced by removing all impurities from the sand. The next step is to make a single long cylindrical crystal of this almost pure silicon, which has been intentionally doped with a small concentration of phosphorus. The phosphorus allows this silicon crystal to conduct negative charges, or electrons, and so is named N-type silicon crystal. The cylinder is then sliced into very thin wafers (0.025cm thick) that become solar cells. Some manufacturers use ignot casting to produce square polycrystalline cells. Their shape allows more of them to be mounted together. Still others are developing techniques to produce solar cell material in continuous sheets or thin films glued to glass panels. Production costs potentially are much lower than single crystals.

After the wafers have been cut, their front surfaces are treated with a gaseous derivative of boron, which diffuses into the wafer surfaces, producing a very thin P-type layer, which means it conducts positive (P) charges. The boundary where the N-type meets the P-type crystal is called the P-N Junction, and is the place where the photovoltaic effect occurs.

When a single photon of light is absorbed at the P-N junction, it creates one negative charge and one corresponding positive charge. The negative charges tend to congregate in the N-type layer and the positive charges in the P-type layer, rather like boys and girls at a high school dance. See Fig. 16-A.

If metal contacts are provided on each side of the solar cell, and joined, a current will flow. The amount of current that will flow in bright sunshine is strictly dependent on the

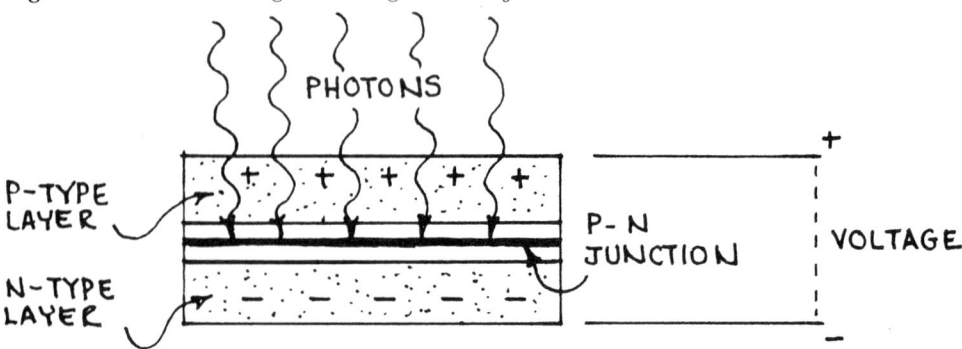

**Fig. 16-A.** Photons of light striking the P-N junction inside a solar cell.

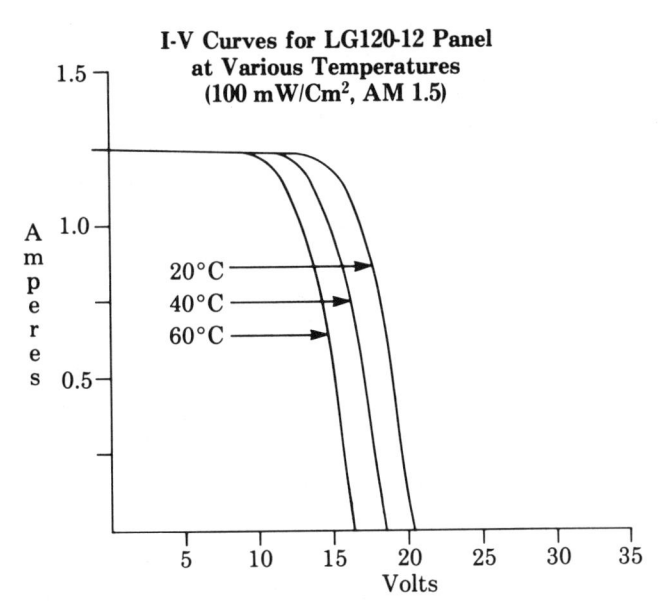

**I-V Curves for LG120-12 Panel
at Various Temperatures
(100 mW/Cm$^2$, AM 1.5)**

Amperes vs Volts, curves at 20°C, 40°C, 60°C

**Environmental Operating Conditions**

Temperature: −40°C to 90°C
Humidity: 0 to 100%
Altitude: to 25000 ft. (7.620m)

Wind Loading: Modules withstand sustained winds in excess of 175 mph (280 kph) or 90 lbs/sq. ft.

**Fig. 16–B.** Supplied by each manufacturer, these curves show how current and voltage will vary with the temperature of a solar cell.

area of the solar cell. NASA defines "peak" sunshine as 100 milliWatts/cm$^2$, and a solar cell with a 100-millimeter diameter (3.94″) can produce more than 2 amps of current. But remember, the power available is equal to current x voltage, and the voltage attained by solar cells regardless of size is approximately 0.5 volts. Therefore, one solar cell 100 millimeters in diameter produces a current of 2 amps at 0.5 volts, or has a power of about 1 watt.

## Solar Panels

Solar panels consist of a group of solar cells connected in series. This means that the *amperage* of the entire panel is the same as that of one cell, but the *voltage* is cumulative and increases with the number of cells. To charge a typical battery you need a higher voltage to overcome the electrical potential of 12–14 volts. Most solar panels connect 36 solar cells in series to produce the necessary voltage. Some panels have more than 36 cells, claiming better efficiency in overcast conditions. Others, like the Arco M-63, have only 30 cells so that one panel connected to one 12V battery will never produce a voltage that can overcharge the battery or require regulation.

The power output of a panel decreases as temperature rises. Cell temperature is a good

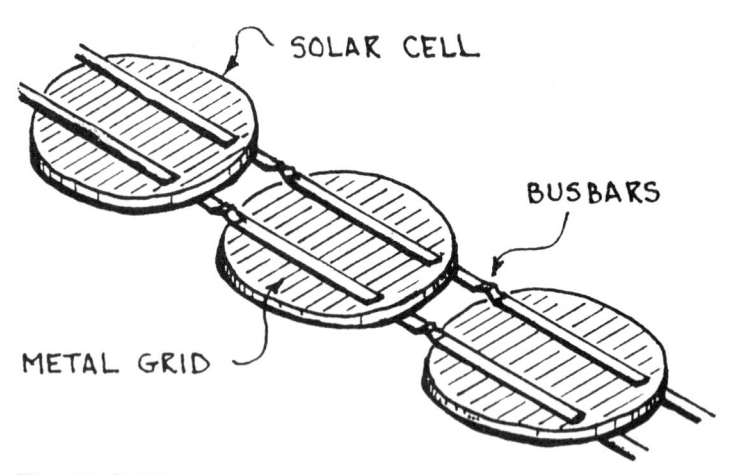

**Fig. 16-C.** The current-carrying metal grid on the surface of a solar cell.

bit higher than ambient air temperature due to the greenhouse effect of the panel's glass cover. Each manufacturer rates its panels at a specific cell temperature and sunlight level, usually 25°C and 100 mW/cm$^2$. For an example of panel performance in other conditions, see Fig. 16-B.

It is important to keep high temperatures in mind when you mount your solar panels so that air is free to circulate underneath and around them.

It's easy to provide a metal contact to the back side of a solar cell to conduct electrical charge, but any metal area on the front will reduce the amount of surface exposed to the sun. For this reason solar panels are provided with a super-thin metal grid on the front of their cells for maximum performance. The grid is strengthened by a pair of conducting bus bars, one on each side of the cell. See Fig. 16-C.

The cells are placed on a light-colored reflective backing sheet (typically Mylar®) and encapsulated in a material that protects them and holds them in place. Encapsulating materials include silicone rubber, PVB (polyvinyl butyral) and EVA (ethylene vinyl acetate). Most manufacturers now prefer EVA because of its ultra-violet resistance and stability with respect to color fading, humidity and temperature.

A glass or polymer cover is placed over the encapsulated cells with a gasketed frame secured to the perimeter. The junction box for the positive and negative contacts of the panel should be well protected from the environment. A good example is shown in Fig. 16-D.

The usable power output from a solar panel varies with the size of the solar cell, and therefore the size of the panel. Typical panels are 7-watt, 10-watt, 20-watt, 30-watt, 35-watt, and 40-watt. These ratings are for peak, or maximum, power. The maximum output you get from a panel will be slightly less than this depending on the amount of sunlight

| Shading | % of Maximum Output | Tilt Angle | % of Maximum Output |
|---|---|---|---|
| Full exposure | 100% | Panel perpendicular (0°) to sun | 100% |
| One thin shadow across panel (as from stay, sheet or halyard) | 95% | Panel tilted 22.5° from sun | 95% |
| Three thin shadows across panel | 85–90% | Panel tilted 45° from sun | 85% |
| One thick shadow 3–5" wide across panel | 50% | Panel tilted 77.5° from sun | 60% |
| One-third panel shaded by object at least 2–3' away from panel | 35–40% | Panel tilted 90° from sun | 30% |
| Two-thirds panel shaded by object at least 2–3' away from panel | 25–30% | Panel tilted more than 90° from sun | 15–20% |
| All of panel shaded by object at least 2–3' away from panel | 15–20% | NA | NA |

and whether it is placed directly perpendicular to the sun's rays. The chart above shows how shading and tilt angle affect photovoltaic panel performance.

## Notes on Solar Panel Output

The effects of shading and the angle of the sun are additive. If a solar panel is tilted 45° to the sun *and* is shaded by three thin lines, the output will be .85 x (.85–.90) = .75, or about 75% of a panel's output when perpendicular to full sun. Contrary to popular belief, a few thin shadows cast by rigging on the boat will not appreciably affect a solar panel's output.

For best overall results throughout the day, simply mount the solar panel horizontally.

Reflected sunlight from the water helps to lighten shadows cast on the panel and may increase solar performance by 20%. A blanket of snow will have the same effect on land-based solar systems. When performing tests on *Kjersti* from which the last table was com-

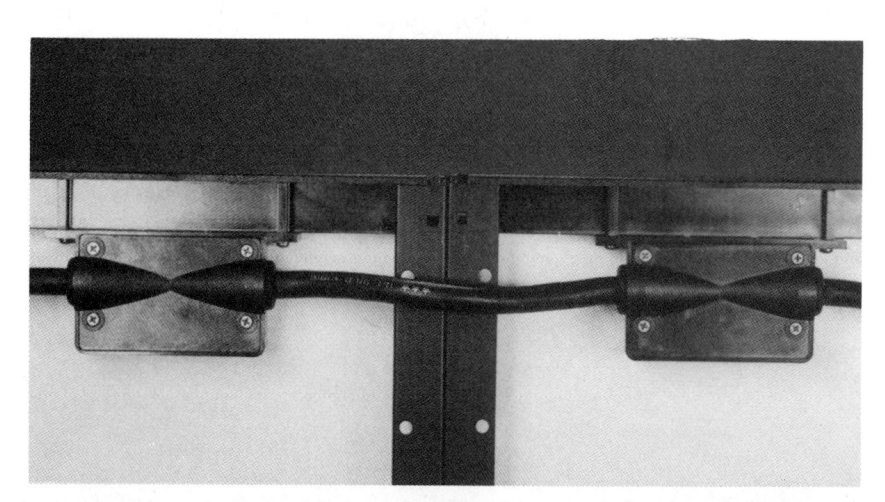

**Fig. 16-D.** A well-designed electrical junction box on the rear of a solar panel (courtesy of SPC).

piled, we discovered glare contributed to the high output readings obtained when the panel was tilted away from the sun.

## WIND GENERATORS

A simple formula exists for describing the amount of energy available to things in motion, including moving air:

$$E_{(k)} = \frac{mv^2}{2}$$

and for the energy in the wind:

$E_{(k)}$ = Kinetic energy of the wind
m = Mass of the air
v = Velocity (speed and direction of the airstream)

This section deals only with propeller-type wind generators. First, let's determine the mass of the air that travels past our generator propeller in a given amount of time. It is expressed as:

m = eAvt where e = the density, or wt./volume, of the air
A = circular area defined by rotating prop.
v = velocity of the wind
t = the time elapsed.

We can substitute this expression for the air mass (m) in the energy equation, which then becomes:

$$E_{(k)} = \frac{(eAvt)v^2}{2} = \frac{eAtv^3}{2}$$

Theoretically, this is the total amount of energy that is available in the wind over the time period t. Since power equals energy divided by time, our formula for instantaneous power becomes:

$$\frac{E_{(k)}}{t} = \frac{eAtv^3}{2t} \text{ or } P = \frac{eAv^3}{2}$$

This is the theoretical maximum power available from the wind before it comes in contact with our propeller. But, as the airstream tries to pass through a wind generator propeller, the air is slowed down from $v_1$ (initial velocity) to $v_2$ (downward velocity). At the same time, the propeller tries to transform most of the horizontal kinetic energy of the wind into rotary kinetic energy (the motion necessary to turn a generator). The propeller's ability to transform this energy is called its efficiency.

Figure 16–E shows what happens as an airstream encounters a propeller. Its area is increased to equal the swept area (A) of the propeller (which must be free to rotate) and has its velocity reduced to $v_2$. Using our knowledge of these events, and the equation for P, above, we can calculate the efficiency of the propeller.

The actual power we extract from the wind is related to the difference between $v_1$ and $v_2$,

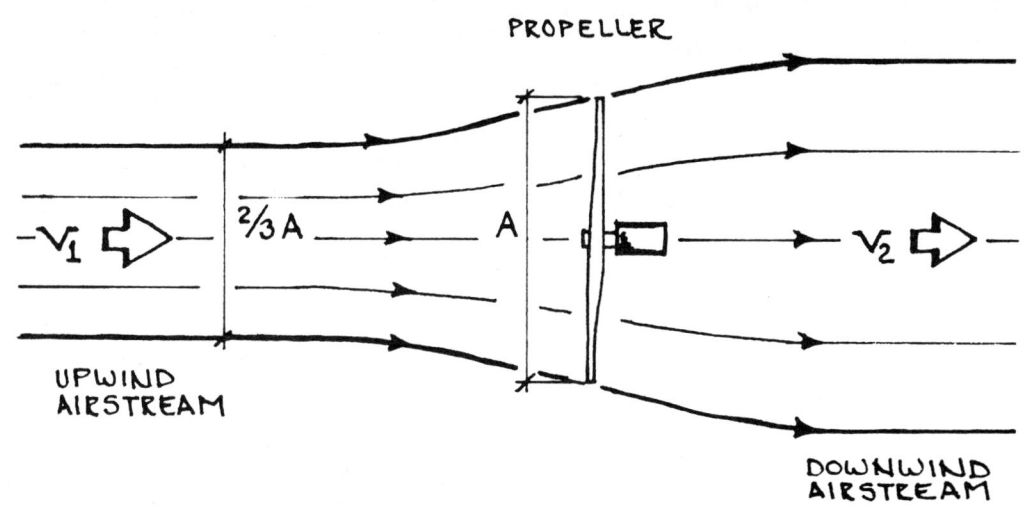

**Fig. 16-E.** Characteristics of an airstream meeting a wind generator.

or how much we slow down the airstream through the propeller. The power increases as $v_2$ decreases until it reaches $\frac{1}{3}$ of $v_1$. At this point the power is at a maximum and reducing $v_2$ also reduces the power. See Fig. 16-F.

So, the maximum theoretical power that propeller designers try to achieve is really only 59.3% of the power of the wind that would pass through unobstructed. Maximum theoretical power is expressed mathematically as:

$$P_{max} = 0.593 \times \frac{eAv^3}{2}$$

This expression is known as Betz's Law. Although certain experimental wind machines have exceeded the maximum power shown here, it is suitable for our purposes.

To make this formula meaningful and useful we need to determine the density of air (at sea level, of course) and be able to express maximum theoretical power ($P_{max}$) in watts, area (A) of propeller diameter (D) in feet, and velocity (v) in miles an hour. A good value for air density is .08 lb./ft$^3$, and the circular area swept by a propeller is $\frac{\pi D^2}{4}$. But trying to get the equations to cancel out is enough to drive you mad. Fortunately, with the help of the authors of the superb alternate energy primer *Other Homes and Garbage* we can make the formula come out nicely as:

$$P_{max} = 0.0024 D^2 v^3 \text{ (theoretical)}$$

Now that the theory is behind us, what happens when we hoist our wind generator into the foretriangle or up on a permanent mount? How much electrical power actually gets to the battery? Our theoretical power decreases by the following:

*Efficiency of Generator (E Gen)*—Most generators or alternators only transform about 70% of the rotary mechanical energy into electrical energy at the battery. That is, we have a loss factor of 0.70.

149
*A Little Theory*

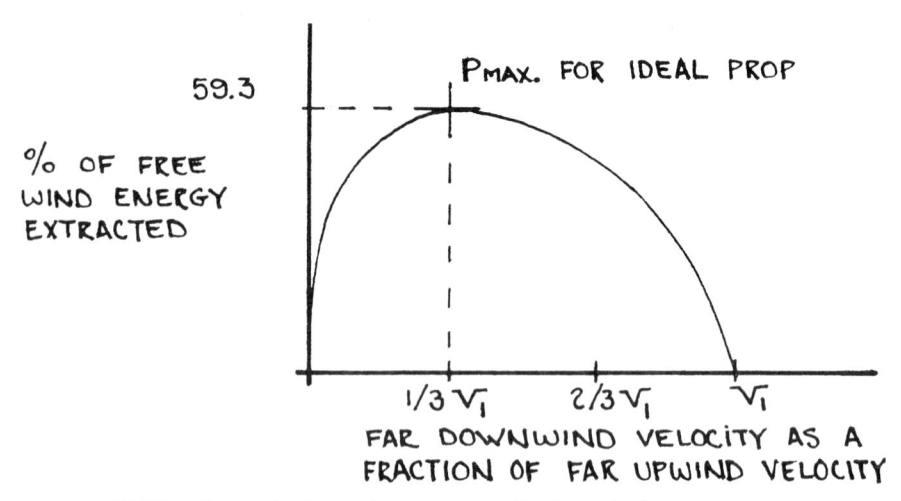

**Fig. 16-F.** The theoretical maximum power in the wind.

*Efficiency of Drive System (E Drive)*—Energy losses can occur between generator and propeller due to gearing, although all but one of the marine generators presently on the market are connected directly to the propeller. The efficiency of a direct drive is 100%, or a factor of 1.0.

*Efficiency of Propeller (E Prop)*—Even a well-engineered propeller only takes advantage of about 60% of the theoretical power available. The factor is 0.60.

The efficiency of the system is equal to E Gen x E Drive x E Prop., or 0.70 x 1.0 x 0.60 = 0.42. This means a good wind generator can extract 42% of the theoretical maximum power available. Therefore, our final expression for usable power is:

$$P = 0.42 \times 0.0024 D^2 v^3 = 0.001 D^2 v^3$$

After all this theory, are we anywhere close to being able to predict the output of a wind generator? Let's see. Let's take a wind unit with a 5' diameter propeller, operating in a 15 mph (just over 13 knot) wind:

$$P(watts) = .001 \times (5)^2 \times (15)^3 = 84 \text{ watts}$$

Using the electrical power formula P = VI and assuming a medium charged battery at 12.5 V, we get:

$$I = P/V = 84W/12.5V = 6.72 \text{ amps}$$

My Hamilton Ferris unit is rated at about 6 amps at 15 mph windspeed, so our calculations are fairly close.

## Notes on Wind Generator Output

System efficiency will vary with different wind generators. Wind generator manufacturers are always trying to find more efficient propeller/generator combinations. The key to increasing efficiency is matching optimum generator or alternator speed to optimum propeller speed. The generators or alternators seen on manufactured units are made to run at low speed. Some are even specially wound and matched with a particular propeller. This is the reason why units with smaller diameter propellers often have a higher system efficiency.

## WATER GENERATORS

To find the energy associated with a water-driven generator, use the same equation for kinetic energy as for wind power:

$$E_{(k)} = \frac{mv^2}{2}$$

$E_{(k)}$ = Kinetic energy of moving water
m = Mass of the water
v = Velocity of the water

When you trail a propeller through the water, it's actually the propeller that's in motion and not the water. Therefore, v actually is the speed of the boat.

And, since water and air are both fluids (yes, air is fluid!) we also can use the same expression for mass as before:

$$m = eAvt$$

e = Density of water
A = Circular area of water prop
v = Velocity of the boat
t = The time elapsed

Unlike air density, water density is not affected by either pressure or temperature.

Since the above expressions for energy and mass apply to wind and water generators, the formula for the maximum power available also is the same:

$$P_{max} = 0.593 \times \frac{eAv^3}{2}$$

When you compare the charging curves for wind and water systems, you see that a water generator, with lower velocities than a wind unit (3–10 knots water vs. 8–25 knots or more wind), and a much smaller propeller (0.75′ vs. 5′) is able to produce as much power as a wind generator. The answer lies in the density of water. Water, with a density of 62.4 lb./cu. ft. (seawater is slightly higher than this), is about 780 times denser than air. In our search for the usable power output of a water generator, we start by assuming that the expression for the maximum theoretical power a water propeller can extract from the water is the same as for our wind propeller (as they are both propellers rotating in fluids). But first we must substitute the water density for air density. From our section on wind power:

$$P_{max} \text{ (wind)} = 0.0024 D^2 v^3$$

Therefore, $P_{max}$ (water) = $0.0024 D^2 v^3 \times \frac{62.4 \text{ e water}}{0.08 \text{ e air}} = 1.87 D^2 v^3$

This is a water generator's theoretical maximum power. However, as with the wind generator, the usable power output of the water generator is somewhat less than its theoretical output. How much less depends on the system efficiency:

*Electrical Generator Efficiency (E Gen)*—We can use the same 70% efficiency (0.70 multiplier) as a wind generator, since most manufacturers use the same generator for both wind and water units, with conversion kits so you can easily switch from wind to water (or vice versa).

*Drive System Efficiency (E Drive)*—Water generators have direct drives. Most water generators are the trailing-log type with approximately 60′ of tightly wound line between generator and propeller. For every revolution of the propeller the shaft revolves once at the other end of the line. For direct drives, the efficiency is 100% (1.0 multiplier).

*Propeller Efficiency (E Prop)*—For most water generators, the propeller is similar to a small outboard propeller mounted on a shaft in reverse. There is little information available to accurately describe propeller efficiency in this appliction. We can find out by consulting a charging curve supplied by a manufacturer. Working backwards we can solve the problem for the missing propeller efficiency "E Prop".

For solving the usable power calculation, let's use a Hamilton Ferris water unit with a propeller diameter of 0.69 ft. and a charging curve as shown in Fig. 16-G.

Our expression for the usable power available from a water generator to charge a battery looks like this:

$P(\text{water}) = 0.70 \times 1.0 \times E\ Prop \times 1.87\ D^2 v^3$
$P(\text{water}) =$
$0.70 \times 1.0 \times E\ Prop \times 1.87 \times 0.48 \times v^3$
Therefore: $\dfrac{P(\text{water})}{0.63 \times v^3} = E\ Prop$

From the information in the charging curve we can select various boat velocities, calculate the actual power available (amps x volts), and solve for the propeller efficiency in each case. The unknown value for the effi-

**Fig. 16-G.** The output curve for our Hamilton Ferris water generator.

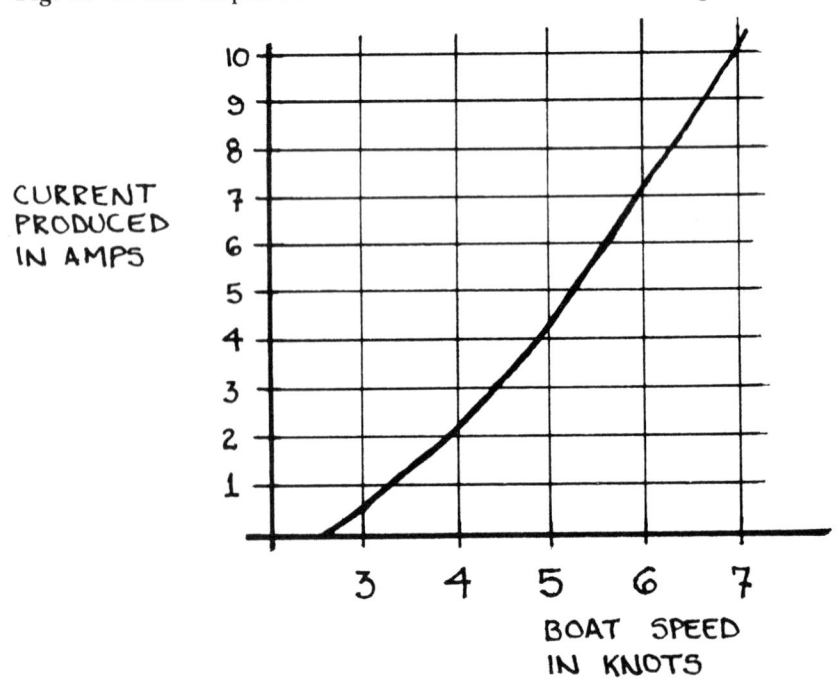

| Boat Speed | | I | Power (watts) | | Prop |
| Kn. | MPH | Amps | Amps x 12.5 Volts | Velocity (v3) | Efficiency |
|---|---|---|---|---|---|
| 4 | 4.6 | 2.0A | 25.0W | 97.33 | 0.40 |
| 5 | 5.76 | 4.5A | 56.25W | 191.1 | 0.47 |
| 6 | 6.91 | 7.4A | 92.5W | 330.0 | 0.44 |
| 7 | 8.06 | 11.0A | 137.5W | 524.0 | 0.42 |

ciency (which is what we want to know) should be about the same for any speed within the boat's normal cruising range, and in fact is, as shown in the chart above:

In the 4–7 knot range, the cruising speed for most boats, the efficiency of the propeller does seem to be relatively constant at about 0.43. As expected, this is much lower than that of a well-designed wind propeller. So, the system's theoretical maximum efficiency is $0.70 \times 1.0 \times 0.43 = 0.30$. The expression for the usable power output of a water generator operating on a boat is:

$$P(watts) = 0.30 \times 1.87\, D^2 v^3 = 0.56\, D^2 v^3$$
D = Diameter of prop in feet
v = Velocity of boat in MPH

Any commercial water generator will be accompanied by a charging curve and power scale defining output at various cruising speeds. If you decide to make your own, this calculation will be close enough to work with, although the efficiencies of the generator and propeller you choose may be quite different. Remember, generators used on all marine alternate energy units are a special type, made to perform at low RPM and very unlike a standard car alternator, for instance.

## Notes on Water Generator Output

Assuming a generator efficiency of 0.70 does not imply that any generator or alternator

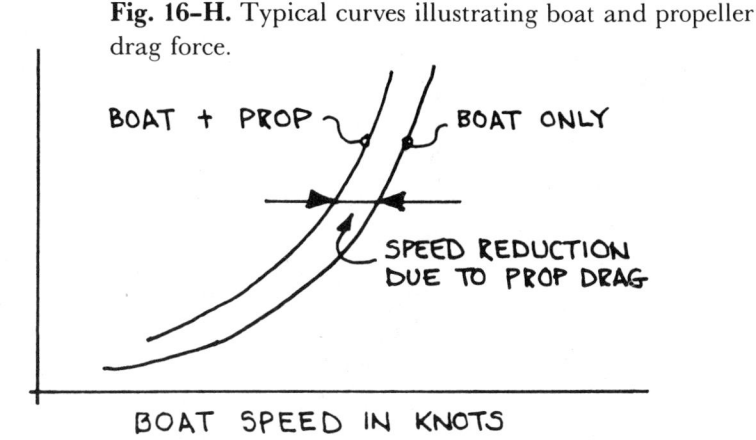

**Fig. 16-H.** Typical curves illustrating boat and propeller drag force.

used in a home-built unit will perform in the same way, or that the efficiency of other manufactured units is necessarily the same. Only if you select a well-made electrical generator that is sized and wound to properly match the RPM available from your propeller, will your maximum efficiency approach 0.70.

If a large free-wheeling or auxiliary propeller through the hull is used instead of a trailing-log type, the possibility exists to gear up the shaft to the higher RPM used by more standard generators/alternators. But you must take into account the loss due to the inefficiency of the gearing system, which might be 10–20%.

Assume that the efficiency of the electrical generator remains constant at various velocities. While this is not completely accurate, it is interesting to note that for the charging curve we used, the system efficiency remained constant for boat speeds of 4–7 knots.

# GENERATORS/ALTERNATORS

With a few exceptions, the actual generator that creates current when its shaft is rotated by a wind or water propeller is a permanent magnet DC generator.

## Permanent Magnet DC Generator

A permanent magnet generator derives its name from the curved, usually ceramic, permanent magnets epoxied to the inside of the generator casing.

A DC generator (see **Fig. 16–I.**) is composed of the following:
1. A stationary housing, or stator (usually a cylinder of steel).
2. A source of magnetic field.
3. A group of wire windings, or armature, in which the current is actually gen-

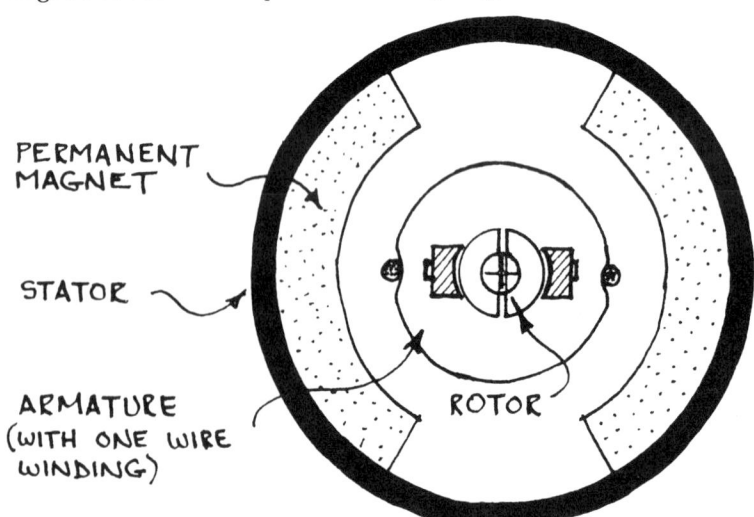

**Fig. 16–I.** Section of a permanent magnet generator.

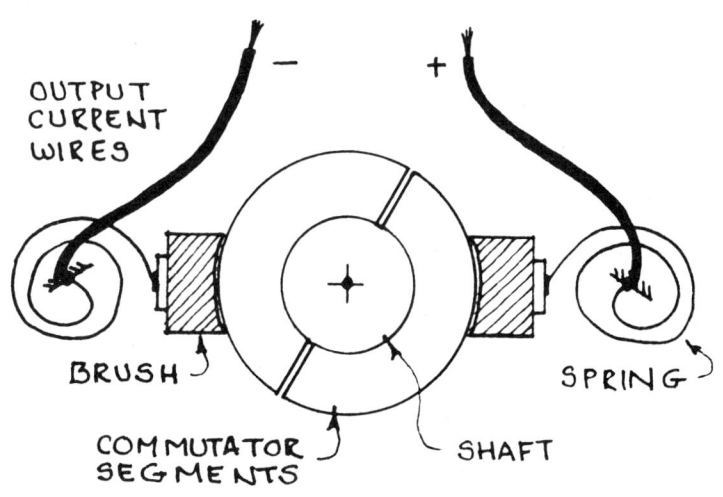

**Fig. 16-J.** Close-up view of the commutator and brushes on a permanent magnet generator.

erated, and which conducts the current to the commutator.
4. A shaft, or rotor, that spins the armature in relation to the motionless magnet and housing.
5. A commutator at the end of the shaft that transfers the current from the rotating winding armature to the external circuitry (see Fig. 16-J). The ends of each wire coil on the armature are connected to a segment of the commutator.
6. Brushes, spring-loaded conductors against which the commutator rotates, accept the current from the rotating commutator and pass it down stationary wires.

Because of the relatively high current being transferred, the commutator segments and brushes take a fair amount of wear, and must be designed accordingly. The brushes, made of a softer material, need replacing at regular intervals (usually after 2 years, possibly more often on a continually used water generator). The voltage output as the shaft is rotated is shown in Fig. 16-K.

Permanent magnet DC generators can produce almost unlimited current and will burn themselves out if the shaft speed (and therefore current) is allowed to increase past a given value. Generator and alternator overheating happens when internal heat produced in the windings cannot be dissipated fast enough.

Resistance can be reduced by using thicker, but bulkier, heavier and more costly windings or by using shorter length of wire, or fewer turns in the windings.

There are, however, three ways to prevent a permanent magnet DC generator from overheating. One method is to electronically control the current with cutout switches that intermittently stop current production in higher windspeeds so the generator can cool

*A Little Theory*

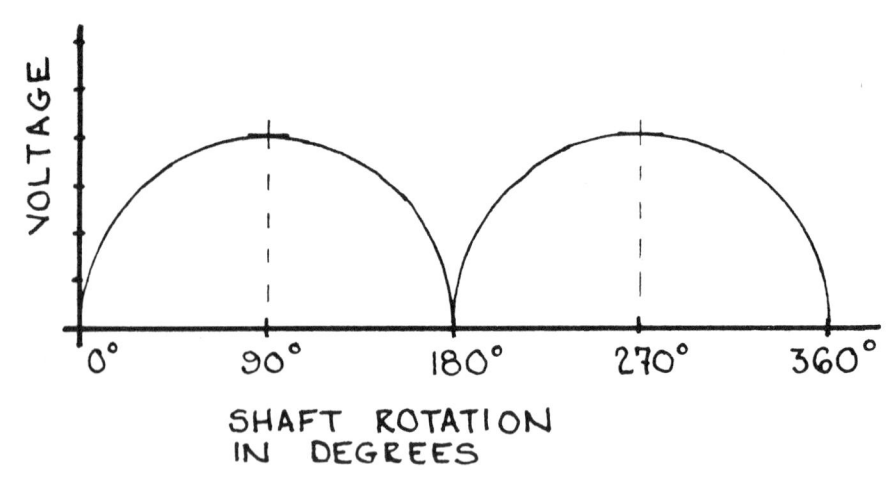

**Fig. 16-K.** Voltage curve for DC generator with 2 magnets and one coil (the simplest case).

down. This is very difficult and expensive to do, because of the arcing problems associated with breaking a DC current.

Another method mechanically governs the speed of the propeller so it will not exceed a safe RPM. This method must be strong and foolproof. Some methods currently in use are spring-loaded air brakes (Winco, Hamilton Ferris), spring-loaded, tilt-up generator/prop assembly (Thermax), and tail vane deflectors that rotate the unit out of the wind (Silentpower).

The least foolproof but most often used method is to tie off or take down the unit when the wind is too strong, usually 25 knots. If you own one of these units that requires manual shutdown, you should realize that this must be done to protect your generator and keep your warranty intact.

## Alternators

An alternator actually is a simplified generator that needs no rotating contacts to carry the output current. By locating the armature (wire windings) in the stator, or stationary housing, the current produced is transferred directly by stationary contacts. The magnetic field is created in one of two ways:

*Permanent Magnet Alternators*—A special type of alternator where large permanent magnets are located on the rotor as shown in Fig. 16-L. This method is employed by Ampair, Powerlog, W.A.S.P. and LVM. Permanent magnet alternators used in wind machines must also be protected from too much current production at high speeds. This is accomplished in one of two ways.

On a permanent magnet alternator with ironless stator, the current can be controlled electronically by temperature-activated cut-out switches. These units spend much of the time at higher windspeeds producing no current, so care must be taken when examining manufacturer output curves, which only give the on-line currents and not the average current.

On a permanent magnet alternator with iron cored stator, winding inductance causes

*Reference Section*

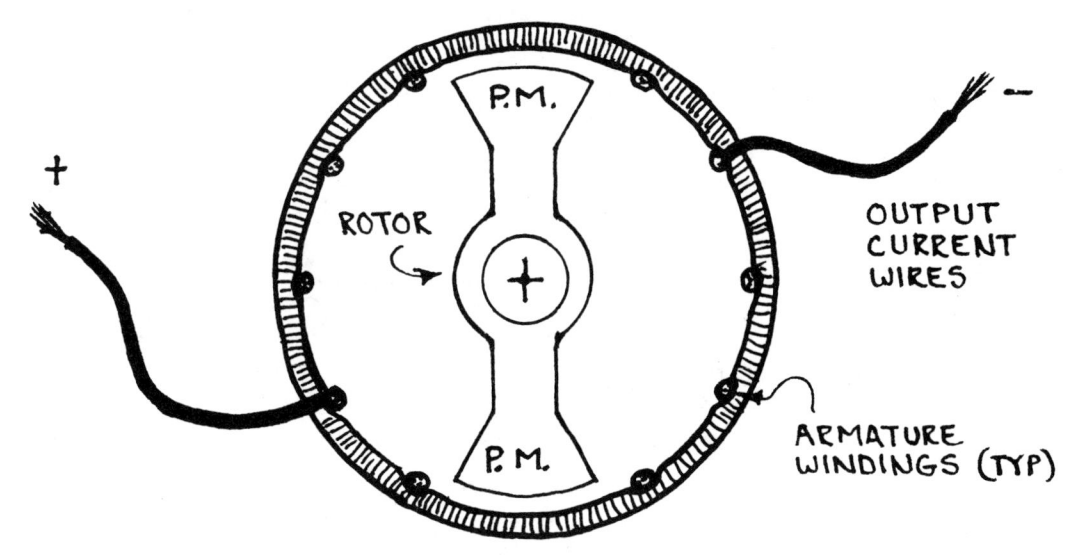

**Fig. 16–L.** Section of a permanent magnet alternator.

**Fig. 16–M.** Section of a field current alternator.

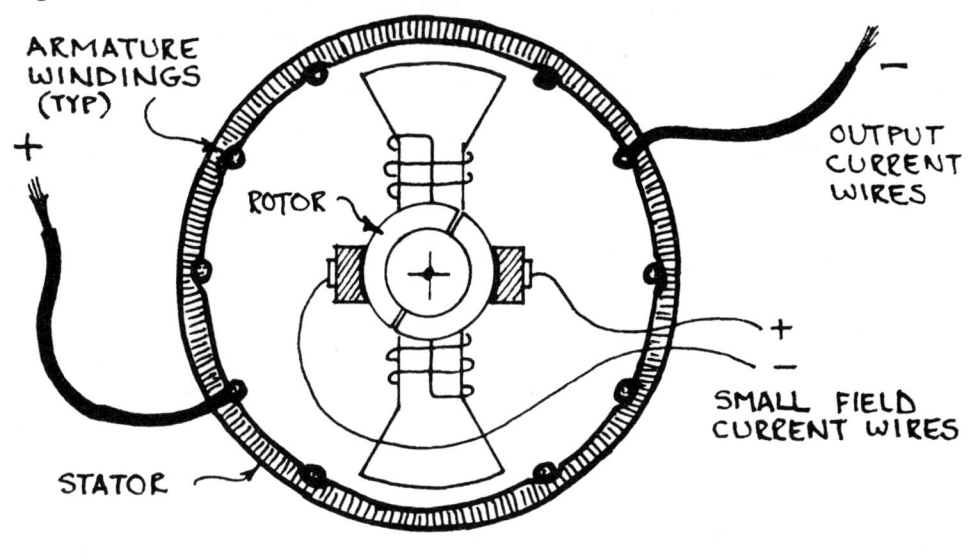

157
*A Little Theory*

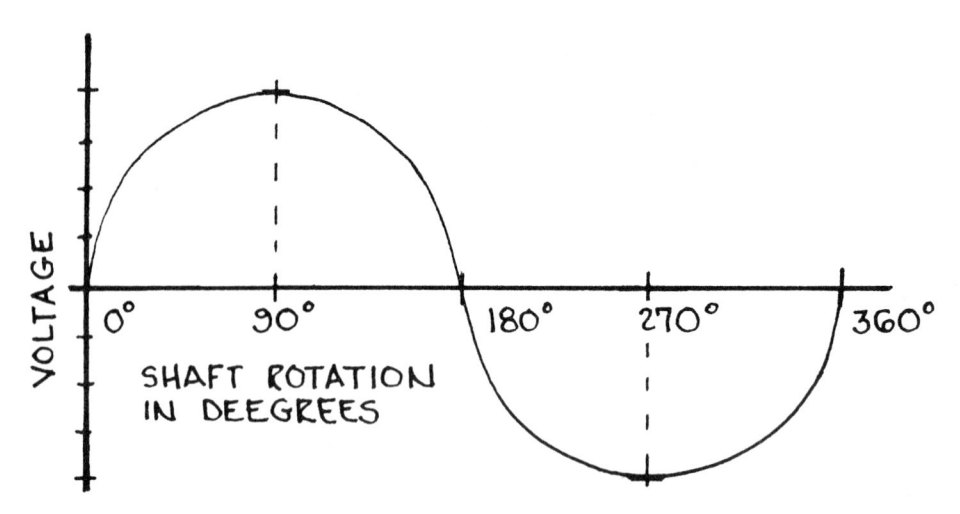

**Fig. 16–N.** Voltage curve for an alternator with 2 permanent magnets and one coil.

the current to remain at a safe value. No switches are required and you maintain full current production at all windspeeds. These are specially-made units that function well at low speeds. Ampair 100 and Aquair 50 use this method.

*Field Current Alternators*—This is the type used for car alternators as well as for several marine wind and water units. As shown in Fig. 16–M, a set of rotating contacts supplies a small electric field current to the wire windings on the rotor that creates the magnetic field. This field current comes from the battery.

The output current from any alternator or generator is AC and increases and decreases with each rotation of the shaft. See Fig. 16–N. Before the energy from an alternator can be stored in a battery, the current must be rectified, or changed to DC. A rectifier is an integral part of most alternators. On a generator the commutator and brushes accomplish the same thing.

# CHAPTER 17
# THE MARKETPLACE

An honest attempt was made in this section to include all alternate energy systems marketed today in the United States. If any product is missing, every attempt will be made to include it in the next edition. A cautionary word is needed on the subject of the listed outputs for the solar, wind and water generators. Because of their large land-based market, solar panels are tested by independent laboratories in strictly controlled conditions. However no such facilities exist for testing marine wind and water generators. There exists, therefore, much confusion about the conditions that resulted in a manufacturer's output listing. Output varies according to battery size and state of charge, temperature of generator (some ratings are for a cold unit and output drops off as unit warms up during operation), ambient air temperature, length and size of battery leads, different windspeed/boatspeed readings, etc. We have tried to temper manufacturers' ratings with observations made during actual use. But listed outputs should only be considered as rough estimates. The obvious solution to these difficulties is to establish an independent testing facility to test and rate marine wind and water generators.

**NOTE:** All prices in this section were in effect as of summer, 1985.

# SOLAR PANELS

## A.E.G. Telefunken/Systems Technology Division
Rt. 22 Orr Dr., Somerville, NJ 08876

This company has been making photovoltaic equipment for more than 20 years, including power supply systems for satellites. Its marine panels were designed by sailors on the engineering staff, at company headquarters in West Germany.

| Model Number | SLB1 (1 Panel) | SLB2N (2 Panels) | SLB2H (2 Panels) | SLB3 (3 Panels) | SLB4 (4 Panels) | PQ5/40/OTS (1 Panel) |
|---|---|---|---|---|---|---|
| Grade | ← | | marine | | | → |
| Module Size (w x l) | 10⅞" x 23⅝" | 21¾" x 23⅝" | 10⅞" x 47¼" | 30⅝" x 23⅝" | 40½" x 23⅝" | 9⅝" x 22" |
| Weight (lbs.) | 10 | 20 | 20 | 30.0 | 40 | 5.5 |
| Peak Current (amps) | 0.54 | 1.08 | 1.08 | 1.62 | 2.16 | 0.54 |
| Open Circuit Voltage | ← | | | | 21.7 (typical) | → |
| Peak Power (watts) | 10 | 20 | 20 | 30 | 40 | 10 |
| Cells Per Molecule | ← | | | | 40 (typical) | → |
| Cell Type | ← | | | | square polycrystalline | → |
| Cover Material | ← | | | | smooth, tempered, low-iron glass | → |
| Frame or Trim Type | ← perimeter and back stainless steel with teak frame → | | | | | stainless steel |
| Warranty | ← | | 5 years against 10% loss or more of original power | | | → |
| Other Products | charge regulators, electrical deck fittings, diodes, batteries and inverters, cabin fan | | | | | |
| Retail Cost | $267 | $519 | $519 | $774 | $1024 | $237 |

| PQ5/40/02 (1 Panel) | PQ50/40/02 (1 Panel) | PQ10/40/02 (1 Panel) |
|---|---|---|
| ←——— standard ———→ |||
| 9⅝" x 22" | 18⅜" x 22¼" | 18⅜" x 42½" |
| 3.3 | 8.5 | 14.6 |
| 0.54 | 1.1 | 2.2 |
| ————————→ |||
| 10 | 20 | 40 |
| ————————→ |||
| ←— stainless steel perimeter only —→ |||
| $200 | $308 | $478 |

161
*The Marketplace*

## Arco Solar, Inc.
21011 Warner Center Ln. P.O. Box 4400
Woodland Hills, CA 91365

Arco is a large company with longtime interest in solar energy and can be accurately described as one of the leaders in the photovoltaic industry.

| Model Number | M63 | 701 | M53 | M82 | SDM-1000 | SRM-1000 | TDRM-1000 | TDM-5300 |
|---|---|---|---|---|---|---|---|---|
| Grade | ← standard → | | | ← marine → | | | | |
| Module Size (w x l) | 12" x | 12" x | 12" x | 12" x 14" | 13" x 16" | 13" x 16" | 13" x 16" | 14" x 50" |
| Weight (lbs.) | 10.2 | 12 | 12.5 | 1.6 | 2 | 2 | 3 | 14 |
| Peak Current (amps) | 2.1 | 2.26 | 2.4 | 0.46 | 0.46 | 0.46 | 0.46 | 2.4 |
| Open Circuit Voltage | 17.9 | 21.7 | 21.7 | 20.6 | 20.6 | 20.6 | 20.6 | 21.7 |
| Peak Power (watts) | 30 | 35 | 43 | 7 | 7 | 7 | 7 | 43 |
| Cells Per | 30 | 36 | 36 | 35 | 35 | 35 | 35 | 36 |
| Cell Type | ← single crystal silicon → | | | | | | | |
| Cover Material | smooth, tempered low-iron glass | | | ← Tedlar® polymer → | | | | glass |
| Frame or Trim Type | 1.4" deep aluminum at perimeter | | | ← none → | | | ← teak → | |
| Warranty | ← 5 years against 10% loss or more of original power → | | | | | | | |
| Other Products | charge regulators, batteries, mounting systems, lights | | | | | | | |
| Retail Cost | $390 | $438 | $559 | $160 | $285 | $315 | $368 | $825 |

SunMate® Kits include Arco solar module, all necessary mounting hardware, duplex cable, charge regulator and diode, fuse, and sealers.

**Free Energy Systems**
Holmes Industrial Park, Mount and Red Hills Rds.
P.O. Box 3030
Lenni, PA 19052

Free Energy Systems is a company that deals exclusively in alternate energy products.

| Model Number | 129 SL | 129 SQ | 127 SL |
|---|---|---|---|
| Grade | ← | marine | → |
| Module Size (w x l) | 8" x 30" x ½" | 15¼" x 15¼" x ½" | UNK. |
| Weight (lbs.) | UNK. | UNK. | UNK. |
| Peak Current (amps) | 0.6 | 0.6 | 0.5 |
| Open Circuit Voltage | 18 | 18 | 16.5 |
| Peak Power (watts) | 10 | 10 | 8.25 |
| Cells Per Moledule | 36 | 36 | 33 |
| Cell Type | ← | single crystal silicon | → |
| Cover Material | ← | textured acrylic surface | → |
| Frame or Trim Type | ← | none | → |
| Warranty | ← | one year | → |
| Other Products | charge regulators, batteries, and fluorescent lights | | |
| Retail Cost of | $296 | $296 | $181 |

The 129 SQ can be mounted directly to a cabin hatch.

**Solar Systems Division**
Kyocera International, Inc., 7 Powder Horn Drive
P.O. Box 4227
Warren, NJ 07060-0227

These solar modules are available through Kyocera's Cybernet Marine Products Division, and also are marketed by the Coast Navigation Marine Catalog, Annapolis, MD.

| Model Number | KMC-1000 | KMC-2200 | KMC-4400 |
|---|---|---|---|
| Grade | ←————— marine —————→ | | |
| Module Size (w x l) | 11¾" x 22" x 1" | 19¾" x 22" x 1" | 19¾" x 39¾" x 1" |
| Weight (lbs.) | 4.5 | 7.5 | 13.5 |
| Peak Current (amps) | 0.61 | 1.33 | 2.67 |
| Open Circuit Voltage | 20 | 20 | 20 |
| Peak Power (watts) | 10 | 22 | 44 |
| Cells Per Module | 36 | 36 | 36 |
| Cell Type | ←————— square polycrystalline —————→ | | |
| Cover Material | ←——— smooth, tempered low-iron glass ———→ | | |
| Frame or Trim Type | low profile teak frame with integrated feet | | |
| Warranty | 5 years against 7% loss or more of original power | | |
| Other Products | charge regulators and optional protective covers | | |
| Retail Cost | $279 | $429 | $699 |

**PDC Labs International, Inc.**
P.O. Box 603, El Secundo, CA 90245

| Model Number | SC 10A | SC 20A |
|---|---|---|
| Grade | ←——— marine ———→ | |
| Module Size (w x l) | 14"x 17.5" | 14"x 32" |
| Weight (lbs.) | UNK. | UNK. |
| Peak Current (amps) | 0.7 | 1.4 |
| Open Circuit Voltage | 16 | 16 |
| Peak Power (watts) | 10 | 20 |
| Cells Per Module | 36 | 36 |
| Cell Type | ← single crystal silicon → | |
| Cover Material | smooth, temperaed low-iron glass | |
| Frame or Trim Type | ←——— none ———→ | |
| Warranty | ←——— one year ———→ | |
| Other Products | battery condition indicator, dual battery adapter, stern rail mounts, meters, regulators | |
| Retail Cost | $284 | $399 |

*The Marketplace*

## Solarex Corporation

1335 Piccard Dr., Rockville, MD 20850

Solarex products are marketed through its subsidiary, Energy Sciences, 16728 Oakmont Ave., Gaithersburg, MD 20877. Ask for the complete catalog of solar panels, appliances, science and hobby products, and related equipment, The Solar Wonder Book.

| Model Number | 1270M | HE 51M | SX10 | SX20 | SX110 | SX120 |
|---|---|---|---|---|---|---|
| Grade | ←—— marine ——→ | | ←—————— standard ——————→ | | | |
| Module Size (w x l) | 14"x 15" | 23"x 23" | 12"x 18" | 18"x 22" | 18"x 42" | 18"x 42" |
| Weight (lbs.) | 7 | 33.5 | 6 | 13 | 25 | 25 |
| Peak Current (amps) | 0.6 | 2.1 | 0.52 | 1.05 | 2.08 | 2.29 |
| Open Circuit Voltage | 13.5 | 20.5 | 17.3 | 17.3 | 17.3 | 17.5 |
| Peak Power (watts) | 9 | 34 | 9 | 18 | 36 | 40 |
| Cells Per Module | 32 | 36 | 40 | 40 | 40 | 40 |
| Cell Type | single crystal silicon | | ←—————— square polycrystalline ——————→ | | | |
| Cover Material | textured acrylic | ←—————— smooth, tempered, low-iron glass ——————→ | | | | |
| Frame or Trim Type | none | stainless steel back/ perimeter | ←—————— anodized aluminum 1.5" ——————→ deep at perimeter | | | |
| Warranty | 5 years against 20% loss or more of original power loss | | | | | |
| Other Products | see Energy Sciences catalog for details on related equipment | | | | | |
| Retail Cost | $212 | $588 | $188 | $285 | $449 | $499 |

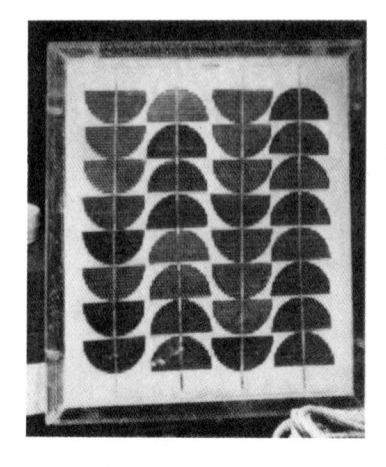

**Fig. 17-A.** Solarex 1270 M marine grade solar panel.

# Solavolt International

P.O. Box 2934, Phoenix, AZ 85062

Solavolt International is affiliated with Motorola Solar Energy, Inc. a subsidiary of Motorola, Inc., and SES, a subsidary of Shell Oil Co. This company is developing a new cell manufacturing process that promises to yield high efficiency (12–14%) cells at a lower cost.

| Model Number | MSP13E10 | MSP23E20 | MSP43E40 |
|---|---|---|---|
| Grade | ←———————— standard ————————→ | | |
| Module Size (w x l) | 14.1″ x 14.9″ | 14.1″ x 21.5″ | 14.1″ x 47.9″ |
| Weight (lbs.) | 6.6 | 9 | 13.5 |
| Peak Current (amps) | 0.55 | 1.1 | 2.3 |
| Open Circuit Voltage | 19.5 | 19.5 | 19.5 |
| Peak Power (watts) | 10 | 20 | 40 |
| Cells Per Module | 33 | 33 | 33 |
| Cell Type | ←———————— single crystal silicon ————————→ | | |
| Cover Material | ←———— smooth, tempered, low-iron glass ————→ | | |
| Frame or Trim Type | anodized aluminum 1.5″ deep at perimeter | | |
| Warranty | | | |
| Other Products | charge regulators, land-based mounting systems, batteries | | |
| Retail Cost | $200 | $350 | $440 |

## Solec International, Inc.
12533 Chadron Ave., Hawthorne, CA 90250

Solec, a member of the Pilkington Group, manufactures and markets solar cells as well as assembled modules.

| Model Number | Solar-Charger I | Solar-Charger II | S-1136 | S-3136 | S-4134 | S-4233 |
|---|---|---|---|---|---|---|
| Grade | ← marine → | | ← standard → | | | |
| Module Size (w x 1) | 13" x 17¼" x ⅝" | 15¾" x 37½" x ⅝" | 12.5" x 15" | 13" x 31" | 15.6" x 37.4" | 12.4" x 47.9" |
| Weight (lbs.) | 2.5 | 9.4 | 3.5 | 6.8 | 12 | 23 |
| Peak Current (amps) | 0.5 | 2 | 0.64 | 1.3 | 2.23 | 4.23 |
| Open Circuit Voltage | 19 | 19 | 20.2 | 19.6 | 20.5 | 20.2 |
| Peak Power (watts) | 9 | 30 | 10 | 20 | 35 | 66 |
| Cells Per Module | 36 | 36 | 36 | 36 | 34 | 66 |
| Cell Type | ← single crystal silicon → | | | | | |
| Cover Material | ← Tedlar® polymer acrylic → | | ← smooth, tempered, low-iron glass → | | | |
| Frame or Trim | ← thin, black edging → | | ← anodized aluminum 1.5" deep at perimeter → | | | |
| Warranty | ← one year → | | ← 5 years → | | | |
| Other Products | charge regulators, batteries | | | | | |
| Retail Cost | $350 | $605 | $185 | $291 | $445 | $745 |

Solec panels are available at reduced cost from Current Alternatives, P.O. Box 166 Northfield, VT 05663. They are the Northeast distributor for Solec, and advertise extensively in *Cruising World* magazine and other sailing publications. Write to Solec for the addresses of other Solec distributors and dealers.

# WIND GENERATORS

## Ampair Products
76 Meadrow, Godalming, Surrey
GU7 3HT, England

Hugh Merewether and Ampair Products have been making wind and water generators for sailors since 1976, and have done much to promote alternative energy on boats.

| Model Name/Number | Aquair 50 | Ampair 100/ Aquair 100 |
|---|---|---|
| Generator Type | ← low speed permanent magnet alternator, self-exciting and self-limiting → | |
| Propeller Type | 26" diameter thermoplastic (14 blades) | 36" diameter thermoplastic (6 blades) |
| Listed Output In 15 Knots Windspeed | 1.25 amps | 2.75 amps |
| Weight (lbs.) | 21 | 26 |
| Mounting Types | permanent rigging suspended | permanent mount, pole* or rigging suspended |
| Generator Housing/Frame | ← aluminum → | |
| Overspeed Governor | ← generators are self-limiting; they do not require shutdown → | |
| Maximum Operational Wind Speed | ← no maximum speed → | |
| Warranty Period | ← one year → | |
| Other Products | voltage regulators, mounting kits, spare parts, water generators, conversion kits | |
| Retail Cost | £250 AQUAIR 50 combined wind & water £350 | £350 AQUAIR 100 combined wind & water £450 |

*Slip rings and thrust bearings included

12V and 24V systems are available. Typical air freight to the United States at the time of this writing is £40.

*The Marketplace*

## Bugger Products, Inc.
P.O. Box 259, Key Largo, FL 33037

| Model Type | Fixed or Pole | Rigging Suspended |
|---|---|---|
| **Generator Type** | ← permanent magnet generator → ||
| **Propeller Type** | 54" diameter 2-bladed wooden propeller ||
| **Listed Output In 15 Knots Windspeed** | 8 amps | 8 amps |
| **Weight (lbs.)** | 30 | 30 |
| **Mounting Types** | fixed, pole* and rigging suspended ||
| **Generator Housing/Frame** | cast aluminum housing, tubular aluminum Lexan tail vane ||
| **Overspeed Governor** | overspeed clutch (mechanical) ||
| **Maximum Operational Wind Speed** | self-limiting with overspeed clutch ||
| **Warranty Period** | ← one year → ||
| **Other Products** | overspeed clutch | |
| **Retail Cost** | $795 | $695 (including mount) |

\* Slip rings and thrust bearings included

Above price includes overspeed clutch, ammeter, diode, and safety shutdown switch.

**Everfair Enterprises, Inc.**
2033 NW 141 St., Miami, FL 33054

|  | ←——————— Fourwinds ———————→ | | | |
|---|---|---|---|---|
| **Model Name/Number** | D54-2 | D54-3 | D72-2 | D72-3 |
| **Generator Type** | ←——————— permanent magnet generator ———————→ | | | |
| **Propeller Type** | 54″ diameter 2 bladed | 54″ diameter 3 bladed | 72″ diameter 2 bladed | 72″ diameter 3 bladed |
|  | ←——————— adjustable pitch wooden blades ———————→ | | | |
| **Listed Output In 15 Knots Windspeed** | 9 amps | 9 amps | 11 amps | 9.1 amps |
| **Weight (lbs.)** | 19 | 20 | 20 | 21 |
|  | ←——————— (does not include mounting assembly) ———————→ | | | |
| **Mounting Types** | fixed, pole* or rigging suspended | | | |
| **Generator Housing/Frame** | no housing; painted generator surface; aluminum frame | | | |
| **Overspeed Governor** | manually controlled electric brake is optional. Automatic electronic brake in development. | | | |
| **Maximum Operational Wind Speed** | ←——————— 25 knots ———————→ | | | |
| **Warranty Period** | ←——————— 3 years from date of purchase ———————→ | | | |
| **Other Products** | long life batteries, voltage regulators, optional mechanical brake, optional control panel, water generators, conversion kits, spare parts, solar panels | | | |
| **Retail Cost** | $569** | $629** | $599** | $679** |

* Does not include slip rings or thrust bearing.

** Above prices do not include control panel ($110 assembled) or rigging mounting system ($89).

A water generator conversion kit is available for $199.

**Friendly Energy Corporation**
P.O. Box 84707, San Diego, CA 92136

| | |
|---|---|
| **Model Number** | 224-R |
| **Gnerator Type** | permanent magnet generator |
| **Propeller Type** | vertical grain Sitka spruce; 2-bladed; 63" diameter glass epoxy insert on leading edge. |
| **Listed Output** | 11.5 amps at 15 knots windspeed (this rating is for a discharged battery) |
| **Weight (lbs.)** | 24 |
| **Mounting Types** | rigging suspended |
| **Generator Housing/Frame** | no housing; anodized aluminum frame castings; fabric tail vane; generator is metal etched, with two coats of primer and two coats urethane |
| **Overspeed Governor** | drum braking device (optional); must be manually reset when activated |
| **Maximum Operational Wind Speed** | approximately 25 knots |
| **Warranty Period** | one year from date of purchase |
| **Other Products** | voltage regulator, water generator conversion kit; protective stowing cases and solar panels |
| **Retail Cost** | $799 |

Does not include control panel. Ammeter and voltmeter can be supplied together for $40.

Water generator conversion kit is available for $160

**Fig. 17-B.** Friendly Energy wind generator.

173
*The Marketplace*

## Hamilton Y. Ferris II Co.
P.O. Box 126 Ashland, MA 01721

Hamilton Ferris II is the nephew of the man who originally designed the Neptune water-powered generator for the 1976 OSTAR. His company now manufactures and markets all types of alternate energy marine products as well as land-based wind and solar systems. Ferris also supplies parts for owner-built generator systems.

| **Model Name** | Neptune Supreme |
| --- | --- |
| **Generator Type** | permanent magnet generator |
| **Propeller Type** | 2-bladed (fir) propeller finished with urethane paint; 60″ diameter |
| **Listed Output** | 7 amps at 15 knots windspeed |
| **Weight (lbs.)** | 23 |
| **Mounting Types** | permanent, fixed pole* or rigging suspended |
| **Generator Housing** | no housing; urethane painted generator; black anodized aluminum frame |
| **Overspeed Governor** | optional overspeed braking device |
| **Maximum Operational Wind Speed** | 25 knots without governor in use |
| **Warranty Period** | one year from date of purchase |
| **Other Products** | water generator, conversion kits; spare parts, voltage regulator, gas and diesel generators, manual alternator control |
| **Retail Cost** | $599—fixed    $699—pole    $709—rigging suspended |

\* Does not include slip rings, but large thrust bearing is included.

Above price includes control panel with ammeter and voltmeter, diode and fuse, wiring, mounting systems.

Water generator conversion kit is available for $219.

## LVM
c/o IMTRA (U.S. distributor)
151 Mystic Ave., Medford, MA 02155

| Model Name/Number | LVM Aerogen 25 | LVM Aerogen 50 |
|---|---|---|
| **Generator Type** | ←——— permanent magnet alternator ———→ ||
| **Propeller Type** | 18″ diameter | 32″ diameter |
| | 5-bladed propellers made of glass-filled polypropylene; mounted on aluminum hub ||
| **Listed Output** | 0.6 amps | 3.0 amps |
| **Weight (lbs.)** | 8.1 | 16.5 |
| **Mounting Types** | permanent mount | pole* |
| **General Housing/Frame** | ←——— cast aluminum ———→ ||
| | takes 1″ tube | takes 1″ tube |
| **Overspeed Governor** | temperature-activated switches self-limit alternator in high winds ||
| **Maximum Operational Wind Speed** | ←——— none ———→ ||
| **Warranty Period** | ←——— not specified ———→ ||
| **Other Products** | LVM 12V tools | |
| **Retail Cost** | $341 | $588 |

\* Slip rings and thrust bearing are included.

Above price includes battery condition indicator and voltage regulator.

## Redwing

Division of JECO Industries, Inc.

Gillespie Airport, Hanger #25, 1935 N. Marshall, El Cajon, CA 92020

Redwing is a leading manufacturer of wind generator kits.

| Model Name/Number | Redwing 48 | Redwing 60 | Redwing 72 |
|---|---|---|---|
| Generator Type | ← permanent magnet generator → | | |
| Propeller Type | 48″ diameter propeller | 60″ diameter propeller | 72″ diameter propeller |
| | ← 2-bladed wooden (fir) propeller with enamel finish → | | |
| Listed Output | 7 amps at 15 knots windspeed for the 60″ diameter propeller | | |
| Weight (lbs.) | ±25 | | |
| Mounting Types | ← rigging suspended → | | |
| Generator Housing/Frame | polished stainless steel housing and frame | | |
| Overspeed Generator | ← none → | | |
| Maximum Operational Wind Speed | ← 25 knots → | | |
| Warranty Period | ← one year from date of purchase → | | |
| Other Products | water generator conversion kit, JECO propellers and wind units for land use | | |
| Retail Cost | basic kit $200 | basic kit $200 complete unit $495 | basic kit $325 |

Basic kits include propeller, generator, mounting hub and diode.

Complete unit does not include control box or meters.

Redwing "Cruise PAC" wind and water generator combination $795

## SeeBreeze Manufacturing
6065 Mission Gorge Rd. Box 120, San Diego, CA 92120

Pat Farrior of SeeBreeze has been building wind generators on and off for the past 8 years. He designs and builds each unit himself. At the time of this writing, he is making a 48″ diameter propeller, and will introduce a 64″ diameter propeller in the near future.

| Model Number | 400-B | 400-BL |
|---|---|---|
| Generator Type | permanent magnet generator | custom permanent magnet brushless alternator |
| Propeller Type | ← 48″ diameter → 2 or 3-bladed hollow-core, pressure laminated epoxy-Kevlar propellers | |
| Listed Output | ← 6.5 amps at 15 knots windspeed (48″ propeller) → | |
| Weight (lbs.) | ← not specified → | |
| Mounting Types | ← rigging suspended or pole mount → | |
| Generator Housing/Frame | ← aircraft-type spun aluminum, aerodynamic hub → | |
| Overspeed Governor | ← none → | |
| Maximum Operational Wind Speed | ← not specified → | |
| Warranty Period | ← not specified → | |
| Other Products | voltage regulator | |
| Retail Cost | $750 for basic halyard mount with 3-bladed propeller | $1000 for basic halyard mount with 3-bladed propeller |

Above prices do not include control panel.

Pivoting assembly for pole mount add $95.

For 64″ diameter blades, add $75.

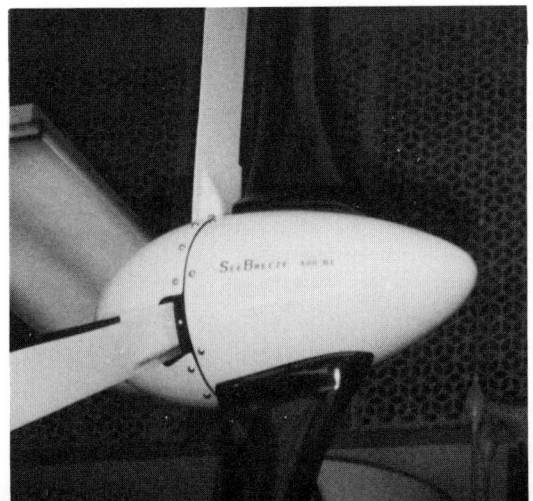

**Fig. 17-C.** SeeBreeze 3-bladed propeller wind generator.

**Thermax Corporation**
One Mill St., Burlington, VT 05401

The Thermax Windstream is primarily a land-based wind system for small homes and cabins, but is adaptable for marine use because of its self tending ability and compact size.

| **Model Name** | Windstream |
|---|---|
| **Generator Type** | permanent magnet generator |
| **Propeller Type** | 2-bladed Sitka spruce 41.5″ diameter; urethane paint finish with elastomer bonded to leading edge |
| **Listed Output** | 2.5 amps at 15 knots |
| **Weight (lbs.)** | 20 |
| **Mounting Types** | pole* mount |
| **Generator Housing** | aluminum housing and frame |
| **Overspeed Governor** | "tilt-up" generator on spring-loaded mechanism |
| **Maximum Operational Wind Speed** | none |
| **Warranty Period** | one year |
| **Other Products** | voltage regulators, inverters, DC motors, control panels |
| **Retail Cost** | $489 |

* Slip rings available as an option.

Price does not include meters, slip rings, or wiring.

## Winco

Division of Dyna Technology, Inc.,
7850 Metro Parkway
Minneapolis, MN 55420

Winco is one of the oldest manufacturers of land-based wind generating systems in this country. Since 1930 it has been providing electricity for farms, rural homes and remote sites. These units sometimes are used on boats even though their weight and steel frame make them most suitable for large ketches (they are usually fix-mounted to the mizzenmast).

| Model Number | W-200-12 | W-450-12 |
|---|---|---|
| **Generator Type** | permanent magnet generator | alternator with rectified 3-phase output (DC) |
| **Propeller Type** | 72" diameter | 96" diameter |
| | 2-bladed wooden (laminated hardwood W-450, non-laminated hardwood W-200) | |
| **Listed Output** | 7.5 amps in 15 knot wind | 12.5 amps in 15 knot wind |
| **Weight (lbs.)** | 60 | 100 |
| **Mounting Types** | boats: fix-mounted to mizzenmast | |
| | land-based: tower mount | |
| **Generator Housing/Frame** | ←—————— painted steel ——————→ | |
| **Overspeed Governor** | ←—————— air brake governor ——————→ | |
| **Maximum Operational Wind Speed** | ←—— none, because of governing device ——→ | |
| **Warranty Period** | ←—— one year from date of purchase ——→ | |
| **Other Products** | voltage regulators, inverters | |
| **Retail Cost** | $795 | $1195 |

These units are available at reduced cost from Hamilton Ferris II Co. (listed in this chapter).

*The Marketplace*

## Wind 'N' Sea Power
5050A Hannah Road, Friday Harbor, WA 98250

The W.A.S.P., as this product is called, is a wind *and* water combination generator, and is marketed as one complete unit.

| | |
|---|---|
| **Model Name** | W.A.S.P. |
| **Generator Type** | permanent magnet alternator |
| **Propeller Type** | 3-bladed hollow epoxy/glass blades, 52" diameter; urethane finish |
| **Listed Output** | 7 amps at 15 knots windspeed<br>3.8 amps at 5 knots boatspeed |
| **Weight (lbs.)** | ±20 |
| **Mounting Types** | rigging suspended, pole (land use only) |
| **Generator Housing** | epoxy/glass housing; resin-treated Douglas fir frame; sailcloth vane |
| **Overspeed Governor** | mechanically feathering propeller |
| **Maximum Operational Wind Speed** | none, because of governing device |
| **Warranty Period** | two years from date of purchase |
| **Other Products** | water generator, land windmill |
| **Retail Cost** | $750 |

Price includes water generator propeller and mount, wind mounting system, and manual brake. Does not include meters or control panel.

**Fig. 17–D.** W.A.S.P. combination wind and water generator.

## Wind Turbine Industries
4035 E. Oceanside Blvd., Oceanside, CA 92056

Steven Gauntt of Wind Turbine Industries previously manufactured and marketed a wind unit with a permanent magnet generator. He switched to a geared drive alternator version that is much more powerful, especially in the higher windspeeds (above 12 knots). He offers a trade-in credit to customers wishing to convert to this system.

| | |
|---|---|
| **Model Name** | Silentpower Supercharger 50 |
| **Generator Type** | self-exciting marine alternator with geared drive to propeller |
| **Propeller Type** | 2-bladed mahogany propeller, 72″ diameter; clear urethane finish |
| **Listed Output** | 16 amps at 15 knots windspeed |
| **Weight (lbs.)** | 30 |
| **Mounting Types** | rigging suspended, pole* |
| **Generator Housing** | polished stainless steel housing and frame, structural black anodized aluminum plates for strength |
| **Overspeed Governor** | a tail vane wind deflection system to feather unit out of the wind is optional |
| **Maximum Operational Wind Speed** | 40 knots without governor, 70 knots with optional tailvane |
| **Warranty Period** | 2 years from date of purchase |
| **Other Products** | small wind unit, control panel, land-use wind system, optional heavy duty manual brake, water genertor, conversion kits, optional tail vane deflector system, other propeller sizes. |
| **Retail Cost** | $1,195 |

* Includes slip rings and thrust bearing.

Price includes wiring, voltage regulator. Water conversion kit available for $200.

## WATER GENERATORS

### Ampair Products
U.S. Distributor: Jack Rabbit Marine, Box 595, Larchmont, NY 10538

| Model Name/Number | Aquair 50* | Aquair 100 |
|---|---|---|
| **Generator Type** | ←—— permanent magnet alternator ——→ (same as wind unit) | |
| **Propeller Type** | ←—— 10″ diameter 2-bladed propeller (black) on 33″ stainless steel shaft and 100′ tow line ——→ | |
| **Listed Output** | 2 amps | 3.5 amps |
| | ←—— rated at 5 knots boatspeed ——→ | |
| **Weight (lbs.)** | 20 | 29 |
| **Mounting Types** | ←—— generator mounted on gimbaled ring located on stern of boat ——→ | |
| **Generator Housing** | ←—————— aluminum ——————→ | |
| **Measured Drag Force** | 20 lbs. at 5 knots | 28 lbs. at 5 knots |
| **Maximum Operational Boat Speed** | 10 knots | 10 knots |
| **Warranty Period** | ←—— one year from date of purchase ——→ | |
| **Other Products** | wind generators, conversion kits, spare parts | |
| **Retail Cost** | £250 | £350 |

Above price does not include postage to U.S.

12V, 24V models available. A break link is included in case propeller snags while underway.

* Just before press time Ampair Products announced this model was being discontinued.

**Fig. 17-E.** Ampair Products' water generator, the Aquair 50 and 100.

## Everfair Enterprises, Inc.
2033 NW 141 St., Miami, FL 33054

| | |
|---|---|
| **Model Number** | WG-1 |
| **Generator Type** | permanent magnet generator (same as wind unit) |
| **Propeller Type** | 9″ diameter 3-bladed outboard motor type propeller mounted on 48″ stainless steel shaft |
| **Listed Output** | 6 amps at 5 knots boatspeed |
| **Weight (lbs.)** | ±20 |
| **Mounting Types** | trailing log type; generator on transom-mounted saddle |
| **Generator Housing** | none, nickel-plated surface with urethane painted finish |
| **Measured Drag Force** | not specified |
| **Maximum Operational Boat Speed** | ±10 knots |
| **Warranty Period** | 3 years from date of purchase |
| **Other Products** | wind generators, voltage regulators, batteries |
| **Retail Cost** | $499 |

Also available is a conversion kit for a free-wheeling propeller shaft to use with existing generator.

**Friendly Energy Corporation**
P.O. Box 84707, San Diego, CA 92136

| | |
|---|---|
| **Generator Type** | permanent magnet generator (same as wind unit) |
| **Propeller Type** | 9″ diameter 3-bladed outboard motor propeller mounted on stainless steel shaft |
| **Listed Output** | 4 amps at 5 knots boatspeed |
| **Weight (lbs.)** | ±20 |
| **Mounting Types** | trailing log type; gimbaled stern mount |
| **Generator Housing** | none, urethane painted finish |
| **Measured Drag Force** | not specified |
| **Maximum Operational Boat Speed** | ±10 knots |
| **Warranty Period** | one year from date of purchase |
| **Other Products** | wind generators, voltage regulators |
| **Retail Cost** | $400 |

Price includes integral thrust bearing.

Friendly Energy is now developing a stern-mounted outboard leg-type water generator.

## Greenwich Corporation
9507 Burwell Rd., Nokesville, VA 22123

| | |
|---|---|
| **Model Name** | Power Log |
| **Generator Type** | permanent magnet (cobalt-samarium) alternator |
| **Propeller Type** | 13″ diameter 2-bladed propeller; high-aspect ratio design; 316 stainless steel |
| **Listed Output** | 5.5 amps at 5 knots boatspeed |
| **Weight (lbs.)** | 15 |
| **Mounting Types** | trailing log type: no mount required as entire unit is trailed in the water; tow line is attached to stern rail or stanchion. |
| **Generator Housing** | black anodized aluminum |
| **Measured Drag Force** | 35 lbs. at 6 knots |
| **Maximum Operational Boat Speed** | not specified |
| **Warranty Period** | one year from date of purchase |
| **Other Products** | none listed |
| Retail Cost | $960 |

Price includes tow cable, mounting bracket, control panel, wiring.

185
*The Marketplace*

## Hamilton Y. Ferris II Co.
P.O. Box 126, Ashland, MA 01721

| | |
|---|---|
| **Model Name** | Neptune Supreme |
| **Generator Type** | permanent magnet generator (same as wind) |
| **Propeller Type** | 3-bladed outboard motor type propeller, 8.25″ diameter Lexan with white finish; mounted on 34″ long stainless steel shaft; 60′ towline |
| **Listed Output** | 4.5 amps at 5 knots boatspeed |
| **Weight (lbs.)** | 22 |
| **Mounting Types** | trailing log type: gimbaled stern-mounted generator |
| **Measured Drag Force** | not specified |
| **Generator Housing/ Frame** | no housing, urethane painted finish; frame is black anodized aluminum |
| **Maximum Operational Boat Speed** | 10 knots without diving plane, 19 knots with diving plane |
| **Warranty Period** | one year from date of purchase |
| **Other Products** | optional diving plane keeps spinner assembly submerged, tested to 19 knots |
| **Retail Cost** | $489 |

Price includes gimbaled mount, tow line, diode and fuse, wiring; does *not* include control panel.

## Redwing Generators

Division of JECO Industries, Inc., Gillespie Airport, Hanger #25 1935 N. Marshall, El Cajon, CA 92020

| | |
|---|---|
| **Generator Type** | permanent magnet generator (same as wind unit) |
| **Propeller Type** | 3-bladed outboard motor type; 9″ diameter mounted on stainless steel shaft |
| **Listed Output** | 3.5 amps at 5 knots boatspeed |
| **Weight (lbs.)** | not specified |
| **Mounting Types** | trailing log type: stern-mounted gimbaled generator |
| **Generator Housing/Frame** | polished stainless steel housing and frame |
| **Measured Drag Force** | not specified |
| **Maximum Operational Boat Speed** | ± 10 knots |
| **Warranty Period** | one year |
| **Other Products** | wind generators |
| **Retail Cost** | $595 |

## Sail Charger Inc.
2895 46th Ave. North, St. Petersburg, FL 33714

This water-driven generator is made to adapt to an existing free-wheeling propeller shaft, and produce electricity while the shaft rotates under sail. This unit can be operated while motoring as well as sailing.

| | |
|---|---|
| **Model Name** | Sail Charger |
| **Generator Type** | geared drive marine alternator |
| **Propeller Type** | can adapt to any free-wheeling type of propeller (will not work with feathering propeller) |
| **Listed Output** | depends on size and configuration of propeller |
| **Weight (lbs.)** | ±15 lbs. |
| **Mounting Types** | auxiliary generator on free-wheeling propeller shaft: drive pulley is mounted on prop shaft; alternator is mounted in line at convenient location near pulley |
| **Generator Housing/Frame** | no housing; frame not specified |
| **Measured Drag Force** | no appreciable additional drag |
| **Maximum Operational Boat Speed** | good for all cruising speeds |
| **Warranty Period** | one year |
| **Other Products** | none specified |
| **Retail Cost** | $895 |

Price includes control panel and voltage regulator.

## Wind Turbine Industries
4035 E. Oceanside Blvd. Oceanside, CA 92056

| | |
|---|---|
| **Generator Type** | geared drive marine alternator (same as wind unit) |
| **Propeller Type** | 8" diameter 3-bladed outboard motor type propeller; 42" long 9/16" diameter stainless steel shaft |
| **Listed Output** | 18–20 amps at 6–7 knots boatspeed |
| **Weight (lbs.)** | 23 |
| **Mounting Types** | trailing log type; stern-mounted swiveling generator |
| **Generator Housing** | polished stainless steel housing |
| **Measured Drag Force** | 35 lbs. at 6 knots boatspeed |
| **Maximum Operational Boat Speed** | ± 10 knots |
| **Warranty Period** | 3 years from date of purchase |
| **Other Products** | wind generators |
| **Retail Cost** | $850 |

Price includes control panel, wiring, voltage regulator. Does not include tow line.

# CHAPTER 18

# COMPARING SOLAR, WIND AND WATER GENERATORS

In Chapter 5 we evaluated what types of AE systems suited your boat. In Chapter 6 we took a look at these systems in light of the kind of sailing you do. In Chapter 7 we evaluated how best to take advantage of your own particular climate in generating electricity.

This final chapter is intended to help you make the choice in selecting your AE system, by comparing various performance and cost criteria for individual products and systems on the market today. Remember while studying the following material that every AE product is right for one sailor or another. Some people may be more concerned about reliability than cost. Others might care more about ease of maintenance and handling, instead of power output. Determine what criteria are most important to you and then judge the various products and systems accordingly.

Whenever possible, evaluations were made of individual products. In some evaluation categories, however, the subject does not lend itself to a product-by-product ranking or critique. In these cases, general comparisons are made between various generic types of AE systems.

## PERFORMANCE

This section will compare how the AE systems perform in the varying conditions encountered while cruising, and what special considerations might be necessary for each system.

### Solar Panels—All Conditions

All panels, regardless of where, when or how you sail, will do some charging during daylight hours, except on the darkest days.

Thin shadows (as from rigging) have a negligible effect on the panel's output. Large shadows will drop the output considerably.

Generally speaking, marine-grade panels take more abuse than standard panels. It is next to impossible to rank individual products by performance since output always will be related to the number of cells in any panel rather than to other differences between products.

Solar panel performance is greatly increased by installing reflectors to direct more sunlight onto the panel surface (see Chapter 12).

## COST AND OUTPUT

The individual prices and electrical outputs were listed in Chapter 17. Divide the cost by the output to get a ratio of cost per amp delivered for ease of comparison. The output current listed was taken from manufacturers' ratings in the following conditions: Solar—bright sunshine; Wind—15 knots; Water—5 knots boatspeed. For an accurate comparison note what accessories are included in the listed price as described in Chapter 17.

|  | *Cost/Amp Delivered* |
|---|---|
| **SOLAR** | |
| Marine-Grade Panels | |
|     7–20 watt | $300–$700 |
|     30 watt or larger | $275–$400 |
| Standard-Grade Panels | |
|     7–10 watt | $250–$400 |
|     30 watt or larger | $175–$250 |
| **WIND** | |
| Small Diameter | |
| Propeller Units | |
|     Low output | $200–$400 |
|     High output | $150–$225 |
| 4′–5′ diameter propeller units | $75–$110 |
| 6′ or larger diameter propeller units | $60–$95 |
| **WATER** | |
| Trailing Log Type | $50–$160 |
| Auxiliary Prop Through Hull | Depends on propeller size and configuration. |
| Auxiliary Generator on Freewheeling Shaft | |

## Wind Generators—Light Wind

*Best*—Begin to operate in 5–6 knots of wind:
    Silentpower, Fourwinds D72-2/D72-3 (assumes blade pitch is adjusted for light winds)

*Very Good*—Begin to operate in 6–7 knots of wind:
    Windbugger, Fourwinds D54-2/D54-3, W.A.S.P.

*Good*—Begin to operate in 7–8 knots of wind:
    Ampair 100, Hamilton Ferris, Redwing, Thermax, Friendly Energy

*Fair*—Begin to generate in over 8 knots of wind:
    LVM25/50, Aquair 50, Wincharger

## Wind Generators—Strong Wind

*Best:* Aquair 50, Ampair 100
    Because of their multibladed design and self-limiting generators these units remain in operation in any conditions.

*Very Good:* Thermax, Wincharger (both models), W.A.S.P., Windbugger
    These units include an overspeed governing device as standard equipment and may be left unattended except in extreme conditions. The W.A.S.P. unit also has a manual brake for rapid shutdown.

*Good:* Hamilton Ferris, Silentpower
    These units offer optional overspeed governing devices that allow the unit to be left unattended except in extreme conditions. The heavy duty disc brake on the Silentpower allows easy manual shutdown in any windspeed.

*Fair:* LVM 25/50, Fourwinds, Friendly Energy, Redwing

LVM units are completely self-tending, but their alternator configuration is such that much of the time in strong winds there is no current being produced (see Alternators, Chapter 16). Fourwinds (any model) feature an optional electric brake but it must be manually operated and is only good up to 40 knots. They are working on an automatic version of this device but it also is only effective up to 40 knots. Friendly Energy's drum braking device must be manually reset after each braking action. Redwing: This unit has no overspeed governor and must be manually shutdown in strong winds.

**Wind Generators—General Notes**

A fixed-mount wind unit that cannot swivel to track the wind independently of the boat will lose much of its efficiency as the boat sails back and forth at anchor, lies to a current or sails off the wind.

A wind unit attached to the upper part of the mast or masthead will consistently have more available wind than one mounted closer to the deck (see Chapter 12).

Large wind generators that are kept up under sail will increase the overall windage of your boat, and must be very well secured.

Pole-mount wind units that have a good thrust bearing for sensitive tracking of the wind will be more efficient than ones without this feature, and will rotate more smoothly with less noise and vibration.

**Water Generators—General Notes**

A large fish probably is less likely to try to eat a Power Log, with its entire generator and steel-reinforced cable in the water, than the steel rod and propeller on a synthetic fiber line used by other trailing log units. Neither one would be especially attactive to munch on. If a fish ate the propeller and rod, however, they could easily be replaced, whereas with the Power Log you would lose the entire unit. Generators operating from an auxiliary shaft or on a freewheeling transmission shaft do not have this problem.

All water generators tend to increase the drag on your boat. This is felt more acutely, both physically and psychologically at slower boat speeds. The Silentpower water generator has such a higher electrical output that it could be operated much less to achieve the same results as other water units. This leaves you free to use it only when the boat is maintaining good speed.

The unit with the least drag at 6 knots is the Aquair 50 (25 lbs.). Most other units rate 35 lbs. at 6 knots.

With the introduction of the Hydro-plane device, the Hamilton Ferris trailing propeller is able to remain submerged at much higher boat speeds than other trailing log units (up to 19 knots). This is especially important to racers and sailors forced to run downwind before the propeller can be retrieved.

**EASE OF OPERATION**

**Solar Panels**

*Best:* Permanently mounted solar panels require no effort for generation. If gimbal mounted or on the stern railing they have the option of being tilted into the sun.

*Very Good:* Moveable solar units are slightly more work than a permanent mount panel. If there is no wind or current charge, you have only to position your panel once or twice a day to catch sufficient sunlight. As our 35W panel produces enough power to satisfy our load, we rarely worry about rearranging its position as the sun moves overhead, unless the panel becomes shaded. Friends of ours with a 10W panel spend more time directing it towards maximum sunlight as they need optimum output, a good reason for oversizing your system.

## Wind Generators

*Best:* Self-tending wind units on poles or on the mast require no shutdown or attention in any conditions. These units include the Ampair 100, Aquair 50, LVM 25/50.

Pole-mounted large-diameter propeller wind units must be tied off in excessive wind speeds if a governing device is not included. They should be taken down in extreme conditions or when sailing long distances offshore, especially downwind. Pole-mounted units with slip rings to transfer current to wires in the pole eliminate the need to periodically untwist the wires.

Fixed large-diameter propeller wind units should have a good governing device and be tied off in extreme conditions. Usually it is shut down but left in place when the boat is under sail.

*Good:* Rigging suspended wind units need to be set up and taken down, as well as tied off in strong winds if a governing device is not included. If you remain at anchor for long periods, the generator can remain in place with little work involved.

## Water Generators

*Best:* Auxiliary propellers through the hull are out of sight, require little or no attention and produce ample power. With through-hull propeller units, the most you have to do is electrically disconnect the system when the batteries are fully charged. A charge regulator eliminates even this small chore.

*Very Good:* Auxiliary generator on freewheeling engine shaft should be electrically and/or mechanically disconnected from the shaft when the engine is running. (This is not necessary with Sail Charger's alternator configuration.)

*Good:* Trailing log water units must be tossed into the water to start, then retrieved when either your boat stops, your battery is fully charged (if no regulator is used), or if you encounter bad weather, excessive speeds or water hazards that might snag the prop. The Power Log is inherently less trouble to retrieve since the towing line does not spin, making it possible to easily and quickly haul it in while underway.

## RELIABILITY/LIFESPAN

Once you've purchased your system, how long will it perform as rated?

### Solar Panels

Some early solar panels had reliability problems that raised doubts about their lifespan. The main problems were corroded electrical contacts or cell surface deterioration, including yellowing or discoloration of the encapsulant material. Another problem was heat applied during metalization and bending of the panel, which caused stress cracking of the solar cell that disrupted the electrical circuit.

These problems have all been eliminated in the latest generation of solar modules from reputable manufacturers. The junction where the wire leads meet the panel is now sealed and the metalization seems to be completely corrosion-proof. A relatively new encapsulant called E.V.A. will not turn brittle and seems to be highly resistant to moisture and ultra violet damage. Better metalization techniques have eliminated stress cracking from heat and glass-covered frames prevent the silicon cells from cracking under normal circumstances. Marine-grade panels protect cells in areas of heavy foot traffic.

Solar panels are said to have a lifespan of 20 years or more, although the cell itself lasts indefinitely. Most solar manufacturers now offer a 5-year warranty against a power loss greater than 10%.

### Wind/Water Generators

All wind and water units are reliable, simple machines with only two moving parts (propeller and generator) and only a slight chance of breakdown. The props will last forever provided they don't hit anything while spinning. On a large-diameter wind propeller the leading edge should be protected against wear from rain, dust and bugs. Epoxy-glass blades are very resistant to wear. Some wood props now have epoxy paint on the whole blade or a thin layer of abrasion-resistant Mylar tape on the leading edge. The generators also will last indefinitely if they are well protected from the environment. The shaft bearings will need periodic replacing, as will the brushes on the DC generators. The Power Log, Aquair and W.A.S.P. units, with their magnetic circuitry, have eliminated all internal rotating electrical contacts, making them potentially very reliable devices. Silentpower's alternator uses rotating contacts only for the small control current, making it a very reliable unit as well. SeeBreeze offers a permanent magnet alternator as an option.

Provided you use them properly, all three systems are extremely reliable with long lifespans and little chance of breakdown.

## SAFETY: A RANKING

What systems pose the fewest potential hazards?

*Best:* Textured or non-skid surfaced solar units have no moving parts of inherent dangers.

Glass-covered solar units are safe if they're out of the way and not regularly walked on. All standard panels and many marine-grade units have a smooth glass cover that can be very slippery when wet.

An auxiliary generator on a free-wheeling transmission shaft is completely safe.

*Very Good:* Small-diameter propeller wind units either mounted on the mast or out of reach on a pole, have a much slower turning speed than the large-diameter props, making them safer to operate and/or handle.

*Good:* Trailing-log water units with 7/16″ diameter braided line and propeller spin at 300–600 RPM, pose a possible hazard to feet, hands or equipment. The boat must come to a stop in order to retrieve the prop unless you use a slit funnel device (see Chapter 13). If you try to pull in your water unit without a funnel while traveling at 3 knots or more, you will get rope burns and a total jumble of line. There is also a remote chance of falling overboard and being hit by the spinning prop, although the longer the line, the less the risk.

The Power Log generator has no inherent dangers. Its line does not spin and it may be retrieved while underway. You would still want to swim clear of its rotating propeller if you went overboard. But you could potentially grab onto the non-spinning tow line. The Aquair 50 and 100 water units have a built in break-link that disconnects the prop from the line if it snags. This is a good safety feature that ought to be standard on all trailing log-water generators.

Auxiliary propellers through the hull are completely safe to operate, although they do require an additional through-hull fitting and stuffing box, which must be installed correctly. This stuffing box will be less prone to leaks, however, since it will not experience the same stresses. But every through-hull fitting carries some risk.

*Fair:* Large-diameter propeller wind units pose a continual hazard if mounted within reach. People have sustained severe cuts from their machines. Properly mounted, they can be accident-proof. The worst offenders appear to be pole and rigging-suspended systems where the owners fail to install them high enough. Raising and lowering the rigging-suspended units or working on and adjusting either type of unit while they are spinning has also been the cause of several accidents. Some systems have also been known to fly apart in high winds, although these accidents all involved homemade frames. However, if you buy a large-prop wind generator that is not self-tending, you must mount it properly and shut it off in high winds. Overspeed governors increase the safety of large wind units, as do heavy duty manual brakes that can stop the propeller in any conditions. In most instances turning the unit 90° to the wind will stop the propeller.

Large propellers painted white with red tips (such as Redwing and Thermax) are much more visible when turning than single-color propellers, and thus are safer. The Friendly Energy propeller, with highly visible orange stripes at the tips, is the most visible stock wind system.

## APPEARANCE

This is not a peripheral issue. Almost any AE system is bound to stand out from a boat's general profile and appearance. But they don't have to stand out like a sore thumb.

*Best:* All water units rate high on appearance. Only the small generator and gimbal for the trailing log can be seen mounted on the stern.

The Power Log is totally immersed except for the towing cable.

*Next Best:* Just about any solar AE system can be mounted unobtrusively on a boat. Generally, the most esthetically pleasing units are those with teak trim followed by thin solar panels with no trim. Among the standard-grade panels, those with a dark colored aluminum frame to match cells are generally more handsome than those with a shiny aluminum frame, although polished aluminum might blend well on boats with a lot of stainless steel fittings and hardware.

*Good:* All large-diameter wind generators are highly visible. Each unit must be rated individually. Small-diameter propeller wind generators are usually mounted high, blend in better with other mast equipment and spin with an unobtrusive motion. Generally, rigging-suspended units are more attractive than pole-mount systems, especially since they can be taken down and stowed when not in use. A sense of esthetics largely is a personal matter.

What follows is a purely subjective critique of the appearance of various AE wind systems.

- Hamilton Ferris and Friendly Energy units are very sleek and good-looking.
- Silentpower and Redwing wind generators feature attractive stainless steel housing and frames.
- SeeBreeze is very attractive, due mainly to its three-bladed design and aerodynamic hub.
- W.A.S.P. features a good-looking, three-bladed design and an attractive fabric tail vane.
- Windbugger has a clean exterior, with its tubular frame it is more attractive mounted on a pole than suspended in the rigging.
- Thermax is clean-looking, small and compact.
- Fourwinds units have a plain exterior. More effort has gone into the engineering than the appearance. The blades are also very different from the airplane propeller appearance of the other large units. But the unit does have a clean and functional look.

## SIZE VS. OUTPUT

Size and weight may or may not be related to the quantity of electricity an AE system delivers. In addition to the issue of output, you must consider what shapes are best for your boat. Let's compare size and weight vs. electrical output in the conditions listed below:

Solar—bright sunshine
Wind—15 knots of wind
Water—5 knots of boatspeed

### Solar Panels

*Size:* Solar panels are very thin ($3/8''$–$1\frac{1}{2}''$), although they present a large surface area; a 35–40W panel's surface area is 4 square feet.

*Weight:* Solar panels are relatively light. A standard 35W panel is 10–13 lbs. Stainless panels weight 40 lbs. for 40W (very tough but heavy). Four large standard panels weigh about 50 lbs., occupy 16 square foot area, and in bright sunshine produce as much electricity as most large-diameter wind units in steady moderate winds.

### Wind/Water Generators

Water probably scores the best among AE systems in terms of size and weight vs. output.

*Size:* The frame, generator and propeller of rigging-suspended mounts requires the maximum amount of storage room, about the same as for a pair of dinghy oars. Pole mounts require no storage, but you may wish to stow the prop periodically. According to owners, the Fourwinds three-bladed prop is very hard to stow without removing two blades from the hub assembly (same applies to the W.A.S.P.). Large-diameter propeller units take up the most room when operating. Trailing-log generators are very compact, use almost no deck space, and are easily stowed. Water generators driven by through-hull shafts or existing propeller shafts take up little space and require no storage.

*Weight:* Both wind and water generators weigh between 20–25 lbs. for each complete unit.

## NOISE/VIBRATION

Compared to even the quietest gas generators or marine auxiliary, AE systems are very quiet. But some do have characteristic noises that are worth mentioning.

### Solar Panels

Absolutely silent!

### Water Generators

Auxiliary through-hull: Propellers sometimes produce a very slight noise as the propeller shaft turns. Usually this sound is indistinguishable from background noise.

Auxiliary generators on free-wheeling shafts can be noisy on certain boats. The shaft should be locked when not motoring or generating.

Trailing logs produce only slight, hardly noticeable flutter or wobble as the line spins.

### Wind Generators

Small-diameter propellers, such as the British units, with their multibladed propellers, are exceptionally well balanced and produce no noticeable noise. In high winds, noise in the rigging is louder than the sound of the propeller. Exception: Pole mounts transfer some noise through the deck, where it is amplified below.

Large diameter wind generators make a slight flutter/chopping noise in higher winds. Exception: Because of its unique airfoil, the Fourwinds has a potentially quieter propeller. Owners adjust the pitch of their own blades, however, so this unit occasionally makes more noise than intended. The amount of noise produced by a given propeller is related to the pitch and airfoil of the individual blades. By increasing the pitch of the blades a manufacturer can increase the efficiency of a propeller in light winds. But at some point, high wind performance is hindered and the blades become noisy as they try to move the surrounding air aside. All wind unit manufacturers try to strike a balance between noise and good overall performance at different windspeeds.

Rigging-suspended mounts isolated by rigging lines (preferably rope), don't transmit noise to other parts of the boat.

Pole-mounted large-diameter propellers transmit much of their noise and vibration below decks. A rubber pad at the base of the pole will lessen, but not eliminate, the problem. A slight shudder also occurs when you change your point of sail. A number of owner-built systems mounted on poles are not properly supported and fastened, and the racket they cause inside the boat is severe.

## MAINTENANCE

What AE systems are easiest to keep in good operating order?

*Best:* Solar panels require no maintenance, other than keeping the cover clean and scratch free and checking the external electrical contacts for corrosion.

*Very Good:* Water units will require some generator maintenance, (see Chapter 13). If a water generator is used continuously during long ocean passages or races, bearings and brushes will require more frequent replacement. Aquair, Power Log, Sail Charger and Silentpower alternators require less maintenance since rotating contacts for output current have been eliminated. The front bearings on trailing-log type water units also are subject to thrust loads from the propeller and line drag that they are not usually designed for. Most bearings used in generators and alternators are only designed to protect the shaft while spinning. This means that periodic bearing replacement is necessary. The Friendly Energy water unit comes with a special bearing for taking the thrust load. The thrust load is less on the Power Log since the propeller and generator are directly connected and there is no line and shaft to create drag on the generator bearing. The Power Log also has an efficiently designed front bearing. A Lexan propeller will not need attention unless it hits something. Metal propellers will need refinishing periodically.

Trailing log systems must be checked for wear on the braded line and for tightness in propeller and shaft connections.

Auxiliary through hull propeller-driven generators require periodic checks for vibration or leaks in the stuffing box.

*Good:* Wind units will require the same general generator maintenance as water units. Painted wooden propeller blades must be refinished more often than expoxy/glass or plastic blades.

## EASE OF REPAIR

What can go wrong and how difficult is it to fix it yourself? Note that this rating category is not the same as a rating for reliability, but how easy or difficult a system is to fix outside of a professional repair shop. Solar panels, for example, are very reliable, but are a problem to fix on the boat, and so they rate lower in this category.

## Wind Generators

Generators are easy to take apart and repair. Usually only bearings or brushes give out. They can be sent to the factory (if under warranty) or repaired at most armature shops. The magnets inside a permanent magnet generator usually are made of a ceramic material that can crack if the generator is dropped. It's possible to epoxy the magnet together until it can be replaced. A good trick to know in an emergency.

Propellers are not easy to fix, due to the critical shape of the airfoil. Most prop/hub assemblies also are factory balanced and if damaged, will probably need rebalancing. On Ampair, LVM and Fourwinds units you can replace individual blades on a prop assembly. Multibladed units (LVM and Ampair) can operate with a damaged prop at lower output if you remove the faulty blade and the opposing blade. With these units it is more feasible to carry spare blades than an entire spare prop. Wood propellers can and should be refinished from time to time.

## Water Generators

Generator repair on water units is similar to wind generators. Power Log alternators have no provision for owner repair. Silentpower uses a standard Motorola marine alternator and therefore repair in remote locations is more likely.

Propeller repair is difficult. Carry a spare prop. A break-link, like the one included in the Aquair unit, is also a good idea so you don't lose your entire generator and gimbaled mount if the prop gets snagged, or pulled by a large fish. Spare props cost about $40–$50—less when used.

## Solar Panels

Solar panel repair is difficult, except for fixing breakdowns in the exterior electrical contacts. Most new panels are factory-encapsulated, which can be damaged by attempting to open, repair and reseal them yourself. While they are more difficult to disassemble, the sealed panels do have a much better record of reliability and long warranty periods.